Sustainable Technology for Society 5.0

This book aims to bring together valuable and novel scientific contributions that address the critical issues of sustainable building, transformative tech models, and other sustainability science and technology topics that have an impact on Society 5.0. This book raises awareness and shares essential policy tools on innovation and technology for sustainable development.

Sustainable Technology for Society 5.0: Case Studies, Examples, and Advanced Research Findings details the use of AI in making complex data analysis and sustainable decision making. It reflects the collaboration of industry, innovation, and infrastructure for Society 5.0. The book elaborates on the essential tools, policy, and strategic implications for building a sustainable tech framework and provides insight into sustainability science and technological contemporary trends. Rounding out the book is a strategic innovative model framework that works towards sustainable, good health, and well-being for Society 5.0.

Researchers, scholars, students, and practitioners will find this book of interest.

Emerging Trends in Technology in Management and Commerce
Series Editors: Vikas Garg and Rashmi Agrawal

In the present competitive scenario the recent changes in hardware, software, and telecommunication technologies has transformed society and is continuously evolving. In the past few decades, there has been tremendous change witnessed in information technology. We have entered the phase where the way we communicate and perform day to day activities is dependent upon telecommunication technology combined with increasing the power of computer use. This change leads to the profound changes in technology that supports our organizations. These changes influence the way organizations function and paves the way for new business opportunities. It also creates the need to understand these technologies in the area of management.

Due to dynamics technology is influencing each and every aspect of an organization. In this regard, the main aim of this new series is to understand the latest technologies that are emerging and changing the entire outlook of management. How the emerging technologies are influencing the communication and day to day activities of an organization. There are a number of books that address different technologies such as AI, blockchain technology, data analytics, multimedia etc. These trends are visible, but at times too complex to be of use to the organization and management team. There is a need for books that can provide the business solutions to many management problems. It is crucial for readers to be exposed not only to different emerging technologies and their applications, but also their impact on management and the role they play in an organization. Strategic managerial issues arising from and dealing with management of information technology and systems are also included in this series.

If you are interested in writing or editing a book for the series or would like more information, please contact Cindy Carelli, cindy.carelli@taylorandfrancis.com.

Sustainable Technology for Society 5.0
Case Studies, Examples, and Advanced Research Findings
Edited by Tilottama Singh, Richa Goel, and Jan Alexa Sotto

Sustainable Technology for Society 5.0

Case Studies, Examples, and Advanced Research Findings

Edited by

Tilottama Singh
Richa Goel
Jan Alexa Sotto

CRC Press
Taylor & Francis Group
Boca Raton London New York

CRC Press is an imprint of the
Taylor & Francis Group, an **informa** business

Designed cover image: Shutterstock

First edition published 2024
by CRC Press
2385 NW Executive Center Drive, Suite 320, Boca Raton FL 33431

and by CRC Press
4 Park Square, Milton Park, Abingdon, Oxon, OX14 4RN

CRC Press is an imprint of Taylor & Francis Group, LLC

ISBN: 978-1-032-42668-6 (hbk)
ISBN: 978-1-032-43052-2 (pbk)
ISBN: 978-1-003-36552-5 (ebk)

DOI: 10.1201/9781003365525

Typeset in Times
by MPS Limited, Dehradun

Contents

Foreword

Sustainable Technology for Society 5.0: Case Studies, Examples, and Advanced Research Findings, which is a part of the "Emerging Trends in Technology in Management in Commerce – Series Editors: Vikas Garg & Rashmi Agarwal", is a comprehensive reference source that will provide personalized, accessible, and well-designed experiences. The aim of the book is to provide a deeper understanding of the relevant aspects of sustainable technology in building an effective Society 5.0. Sustainability has increasingly become important to business research and practice over the past decades because of the rapid depletion of natural resources and concerns over wealth disparity and corporate social responsibility. The book aims at bringing together valuable and novel scientific contributions that address the critical issues of sustainable and transformative tech models and other sustainability science and technology topics that have an impact on Society 5.0 with the fast-changing research community in academia and industry. This book raises awareness and shares essential policy tools on innovation and technology to sustainable development, the possibilities and difficulties of sustainable development, and policy suggestions, especially within the context of the 2030 Agenda towards building a strong sustainable tech framework in the Society 5.0 model.

This aspect has inadequate research and curricular output as of now and could be very useful and a new contribution to this contemporary topic where a customized tech model in different sectors can be adopted leading to sustainable production and profitability patterns for many organizations that pursue sustainability and share the vision of a cleaner planet for the future of mankind.

The book promises that a wide range of readers with a variety of interests will find this book to be extremely valuable. This includes not only academics, postgraduate students, and research associates, but also banking professionals, financial companies, corporate executives, entrepreneurs, and other professionals and masses in all fields who can improve and expand their knowledge by learning the basic trends and activities in this book. Sustainable technology is the need of the hour and can revolutionize any industry as and when needed. As a result, in the coming years, it will affect every single individual on the globe.

This book will showcase these phenomena and literature in the emerging economy context, where innovation and technology are extensively used to achieve the triple bottom line and aid in achieving the SDGs.

Professor Dharam Buddhi
Vice Chancellor, Uttaranchal University
Dehradun, India

Preface

Sustainable Technology for Society 5.0: Case Studies, Examples, and Advanced Research Findings, which is a part of the "Emerging Trends in Technology in Management in Commerce – Series Editors: Vikas Garg & Rashmi Agarwal".

Sustainable technology that is applicable to many industries in the current competitive and technologically driven society has gained paramount importance in the past few years. Building sustainable technology faces more challenging data management, identity theft, and fraud challenges as transactions and other company processes gradually shift online and gain popularity each year. As systems using deep learning technology can detect patterns and spot suspicious activity and probable fraud, building resilient, sustainable tech models in organizations can advance many financial and corporate activities.

With the main objective of designing standard, dependable tech control methods and the search for new ways to reach and serve customers while maintaining low cost, technology-driven organizations have been used in the e-commerce and financial industries to achieve better customer experience, efficient supply chain management, improved operational efficiency, and reduced mate size. Currently, machine learning models are being created to handle the complexity and diversity of data seen in the Society 5.0 Applications of artificial intelligence, machine learning in business, e-commerce, tech management, and tech frameworks are covered in this book. Some of the key uses are portfolio management, fraud detection, inventory management, sales forecasting, profit maximization, and sales growth.

The most recent results of the field's empirical study are presented in this book, along with important theoretical frameworks. It reveals novel and cutting-edge parts of technology applications, demonstrates how it may support sustainable business to increase economic efficiency at both the micro and macro levels, and offers a deeper comprehension of the pertinent facets of technology that have an impact on efficacy for greater output. It is an ideal resource for researchers, academicians, policymakers, business professionals, companies, and students. Numerous practical aspects of artificial intelligence that enhance industry skills as well as decision-making are gaining momentum.

This book is a significant advancement. The book's theme is quite cross-disciplinary in nature. Despite being primarily concerned with stakeholder strategies, the book will be extremely helpful to those in corporate, business professionals, financial markets, the e-commerce industry, sociology, political science, public administration, mass media and communication, information systems, development studies, and business studies. The topic is one of the most significant expanding fields worldwide, and the models covered in the book will have tremendous replication and practice potential. However, for practitioners engaged in the study of stakeholders and their strategies, this book will be a valuable source of reference. Second, the book is organized in a reader-friendly way with key information that has been properly analyzed underlined, making it simple to understand the content. The reader's access to materials in the book creates the possibility for more in-depth research. The case studies will offer a tried-and-true method for resolving typical issues in the subject area. The reader will be able to quickly grasp the chapters' major ideas and summarize the content.

It talks about the enlisted chapters:

Chapter I talks about the impact of technology transformation on sustainable research pedagogy.

The main goal of this research is to examine the blend of learning and digital technology, which talks about systematic planning on research pedagogy of the present learning system. It summarizes some theoretical issues and perspectives for enhancing the effectiveness of integrating digital technology into a scientific and well-advanced mode of the teaching-learning process as related to the new education policy (NEP) in India.

Chapter II talks about enhancing the Impact of Society 5.0 for better healthcare along with social security for the achievement of sustainability.

The goal of all governments across the world is to provide high-quality medical care to every citizen. The experts' recommendations for effective government aid to individuals include promoting healthcare coverage, ensuring a stable foundation, and promoting mindfulness about medical issues. Similar efforts are occasionally made by the Indian government. The Indian government implements several health programmes in India to guarantee people's continued health. With the integration of innovative technologies in every field, we can support the United Nations' Sustainable Development Goals 3 by providing better health for every individual.

Chapter III talks about the significance of smart infrastructure: The role of information technology in smart cities.

The most defining characteristics of the term "smart city" and its development are briefly discussed in this study. Smart cities are those metropolises that incorporate Information and Communication Technologies (ICT) for urban management with the purpose of migrating them from the traditional to the digital and thus improving the efficiency of operations and the provision of services. Along with this, a number of alternative terms that were put forth to describe the many traits of the future cities are also examined. Also provided is a link between technology and smart cities. A smart city is one that develops integrated, liveable, and sustainable urban centres by utilizing all available technology and resources in an intelligent and coordinated way.

Chapter IV talks about chatbots communication quality impact on brand Experience of Customers: An Interdisciplinary Study.

The aspects that may impact brand experiences with the communication quality of chatbots are inadequately known. This study aims to investigate the impact of the credibility, openness, and capability of chatbots communication on the brand experiences of customers. The data that were gathered from 221 respondents were examined using PLS-SEM. The findings of structural equation modelling (SEM) show the significant influence of communication quality variables (credibility, openness, capability) on the brand experiences of customers. This study widens the range of factors that academics and practitioners can consider to speed up the development of chatbot applications beyond user satisfaction and intention to use.

Chapter V talks about role of innovative technology through the study on Tata Nano.

It talks about how the idea of "frugal innovation" is growing in India, where contemporary production methods are adopting it more and more. Despite being rather widespread in India, low-cost innovation has recently become more well-known and accepted. Frugal innovation has become more commonplace and now impacts both urban and suburban communities, in addition to being a mainstay of village and rural life. When it was introduced to the globe, the $2,000 Tata Nano stood as a shining illustration of India's promise of "frugal innovation." Through the use of a case study, the "Tata Nano" project from Tata Motors Ltd. was investigated by means of its cost-effective engineering and design, as well as many enhancements made by the Tata group to integrate their philosophy: "less is more" into the car. The end result, dubbed the "people's automobile," has attracted much interest and acclaim for its creative, economical design.

Chapter VI talks about the role of artificial intelligence in e-commerce: Industry 4.0.

The chapter throws light on Industry 4.0 is revolutionizing, how businesses create, improve, and distribute their goods. Modern technology is being incorporated by manufacturers. i.e., artificial intelligence (AI) and analytics. As an outcome, the chapter highlights how big the sectors can be with AI. The industries can significantly reduce labour costs and complex manual tasks. And they can increase the efficiency and effectiveness of production. They can successfully manage all the sectors of any industry, from primary to banking sectors, for the best of results.

Chapter VII talks about sustainability amalgamation with tech advancement. The study on Safepad: Reusable sanitary pads.

It talks about how women using reusable menstrual products have a better effect on their health. They feel less irritation, burning, or stinging. However, handling these reusable products is comparatively a hassle, as they must be washed and stored in a clean place. Bangladeshi women and young girls still face many social stigmas regarding menstruation, and Youth's Voice aims to spread knowledge and awareness about menstruation. For the betterment of society, it has launched the Safepad Project, which aims to provide a sustainable and healthier option for menstrual products. This campaign aims to reduce the stigma associated with menstruation, remove common myths, inform everyone about menstruation, and distribute sanitary goods to adolescent females in need who are menstruating. The convenience of menstrual products is important, especially the ones that are sustainable and have health benefits.

For women to switch to reusable options, they need to understand the importance of it. Thus, spreading more knowledge and awareness is highly required.

Chapter VIII talks about sustainability through transformative technologies: Green banking and SDG-13.

The current scenario of environmental degradation is enforcing every individual to focus on sustainability. Sustainability has become an increasingly common term in the rhetoric surrounding business ethics and has been widely used by corporations and government consultants pressure drops and academic alike. The study focuses on Ggeen banking, its impact on the industries operating in India, and how technologies have affected the overall productivity of Indian enterprises with their approach to achieving SDG-13. Green banking aims to take proactive action towards environmental protection. The banking industry has a vital impact on society as this can impact our day-to-day life. The current study will focus on green banking as a sustainable contributory solution for the future environment. After facing many environmental disasters, every individual wants to contribute to changing scenarios. Every organization runs its business for profit-making, and hence it is necessary for an organization to know its profit before adopting any changes for sustainability.

Chapter IX talks about managing employees in private organizations: Dilemmas and strategies for managers.

In this chapter, the researcher highlights the challenging tasks of managing employees is a very complex task as they have various order needs. This chapter discusses the issues related with managing employees in private organizations. Workplaces have a combination of personalities, and this factor creates various types of dilemmas. Strategies to deal with the dilemmas are very essential to ensure that every employee maintains focus on the organizational goals. The emphasis is on the private sector employees as they have various insecurities and managing them is a tough task for middle-level managers. With the help of various cases, this chapter helps to point out that with the application of various strategies how managers can cope up with situation that give rise to dilemma. Neglecting the need for such strategies will result in the loss of goodwill of the organization along with higher attrition rate that is not at all favourable for the organization's success. Therefore, any kind of dilemma should be managed timely.

Chapter X talks about the triple bottom line framework and sustainable practices.

The goal of this study is to close the framework comprising the 3 P's: profit, people, and the planet. Leading Indian businesses are increasingly aware of and committed to taking steps to decrease their carbon footprints, such as using materials that were sourced in an ethical manner, cutting back on energy use, and improving shipping efficiency. The study has indeed been able to depict how well the Triple Bottom Line model is implemented in most of the leading companies in India along with other Multinational companies from a global perspective, with a creative touch sensing to social and environmental sustainability. Selection of top innovative Indian companies was identified from various sectors randomly, keeping in mind their contributions towards profits, people, and planet orientations under TBL Scheme for Financial Year 2021. The latter selection of global companies was based on the performance, brand loyalty, customer satisfaction, and corporate image of such companies during the past few years.

Chapter XI talks about the impact of advertising on shopping behaviour.

Consumer behaviour (CB) is highly influenced when any market investor promotes their products through different media outlets. The ad throughout the media has a unique design that involves customers in a different practice. Digital inclusion has prompted changes in buyer media practices. Therefore, a more profound comprehension of promotions on various media stages and the impact on CB should be assembled. Advertising efforts are targeted at the consumer and are in control of everything available to the public. Spending in India is expected to improve by 12.5% in 2018 from 9.6% last year. The present research focuses on a variety of ad factors that affect the performance of each purchase. The people interviewed in this study were college students. It is a well-known fact that the youth of our country are the biggest buyers of smart phones. The conclusion of this study is based on key data collected from students at various colleges in Delhi NCR.

Chapter XII talks about the importance of cost and financial accounting in high-technology firms.

Managers in any industry, including those in the high-technology, are tasked with making tough calls about the financial health of their companies. Financial goals and constraints play a significant role in the day-to-day operations of a high-tech corporation. Expense and revenue accounting are discussed from a general perspective. Due to the breadth of accounting, this will introduce a few of the most important cost and financial accounting concepts as a first step toward recognizing the significance of accounting for management decision in high-technology firms. A system for cost and financial management can boost the automation and digitization of financial management while decreasing the amount of human work performed by financial professionals. Applying this technology in the real world can speed up the input of financial data and enhance the accuracy of accounting records. High-tech companies' financial management systems are highly effective and convenient management solution that meets the majority of their regular business requirements. The system was tailored to the needs of cutting-edge corporations, which is why it excels in this sector.

Chapter XIII provides insights into Kaizen: A Philosophy for Survival and Revival of SMEs after COVID-19.

Taking the theoretical framework from world war II, where Japan was devastated badly and after the world war it was believed that the loss was fairly known to the world, many manifested that the fall of Imperial Japan and believed the loss would be irrecoverable, which surely would have been true in case of any other country, but soon after the trauma, Japan started bouncing back, and today Japan is famous for its Quality, Delivery, and Price, especially one giant automobile company named Toyota, which has set umpteen examples in the field of total quality management. Many researchers have done all-encompassing study and research to find out the magic formula. They found that "Kaizen philosophy" is the backbone and plays a pivotal role in the success of Toyota. This chapter tries to explore the Kaizen philosophy not to survive only but creating a culture of self-sustained continuous improvement in difficult economic scenario after COVID-19, as a number of companies had to be shut down due to losing its customers or suppliers that has created immense pressure on manufacturing industries to work with limited resources.

Chapter XIV instigates the study on Potential Electricity Production of Roof-Mounted Solar PV Systems in a Row House Area in Sweden.

Sweden's energy policy goal is to have completely renewable electricity production by the year 2040; thus, alternatives such as wind and solar energies are being investigated for electricity supply, where mainly solar energy has the potential for small- and medium-sized systems for houses of private individuals and companies. As a part of this development, the tenant-owner's association, Stenbär, in the city of Gävle in Sweden has considered installing solar PV systems, which has become the basis for this research. The aim of this chapter is to investigate the potential areas for a solar cell plant and how much electricity this could produce per year. The simulations are performed by using the computer software IDA ICE 5.0 beta. It is demonstrated that there is a good potential to produce a large amount of solar energy in the area. As the system produces a surplus during the summer half of the year, there is also the possibility of selling the electricity.

Chapter XV talks about the significance of AI in Industrial Revolution 4.0.

Fourth Industrial Revolution has transformed the economies at the global level. New technologies like artificial intelligence have the potential to boost economic growth. Advanced economies have reaped the benefits of technological advancements like artificial intelligence by attracting more investment from poor countries. The chapter aims to explore how artificial intelligence and Industry 4.0 play the role of a double-edged sword in economic development. The chapter discusses how artificial intelligence, the main driving force for Industrial Revolution 4.0, is impacting the economies in general and identifies the factors contributing to both widening and bridging the economic divide. Technological risk in the short and long term is also highlighted, along with the severity of various risks at the global level and insights into the technology-wise impact of Industrial Revolution 4.0 on various organizations, especially the workers. The contribution of the countries in bridging and widening the global economic divide is also mentioned. The chapter leaves the scope for further research in this area.

Chapter XVI talks about the importance of digitalization as a tool for bringing developments and disputations.

AI-powered digitalization in the healthcare industry can be used to evaluate medical pictures, support diagnosis and treatment, and enhance patient outcomes. Similarly, to this, digitalization in finance can be used to spot fraudulent transactions, control risk, and improve investment choices. AI-driven digitalization in the transportation sector can be applied to traffic control, logistics optimization, and the creation of autonomous vehicles. Digitalization is also used in manufacturing to automate monotonous operations, boost productivity, and cut costs. While digitalization offers numerous prospects for increased production and efficiency, it also has drawbacks. The possible loss of jobs as automation rises is one of the key worries. The use of AI raises ethical issues as well, such as the possibility of prejudice and the requirement for accountability and transparency. To ensure that AI serves the interests of society, it is our duty as academicians and practitioners in this area to develop and deploy it responsibly and deliberately. The purpose of the chapter is to shed light on the potential and difficulties brought about by digitalization, as well as the significance of resolving any ethical issues that may arise.

Thus, this book intends to give a quality publication with unique insights and methods of application for current scholars and users. This book offers a great overview of how Sustainable Technology can be built for the futuristic Society 5.0. Its comprehensive coverage paves an interesting read for the users and provides extensive coverage in implementing strategies for building sustainable tech organizations.

Acknowledgement

First and foremost, we would like to thank our All-Powerful God. While putting this book together, we were more aware of how real this writing skill is. You gave us the courage to believe in our passions and pursue our goals. Without your faith in us, we would not have succeeded. We owe a debt of appreciation to our family members, who have stood with us as rock-solid supporters as we struggled to complete this book.

The editors want to thank everyone who helped with this project, especially the writers and reviewers who took part in the evaluation process. Without their assistance, this book would not have been possible. The editors would first and foremost like to thank each contributor and be grateful to the chapter writers who kindly offered their time and talents to this publication. The reviewers made a significant contribution to the quality, coherence, and organization of the information in the chapters, for which the editors are grateful.

The fact that many of the writers also served as referees is something we much appreciate. Those who offered thorough and insightful criticism on a few chapters prompted us to make clarifications, look into particular insight work components, and provide justifications for specific proposals. We'd also like to extend our gratitude to all the folks who have assisted us over the years in learning about and applying the science and art of networking.

Tilottama Singh, Richa Goel, Jan Alexa Sotto

About the Editors

Dr. Tilottama Singh is a certified HR analyst from IIM and a proficient academic, researcher, and trainer with more than 14 years of experience in the field of Human Resources and Work Dynamics. Presently working as Associate Professor and Head of Department in Uttaranchal Institute of Management, Uttaranchal University, Dehradun. Her areas of expertise are Human Resources & Strategy, and her teaching concentration includes Human Resources, Strategy, and Law. She started her career in the hotel industry with Leela Kempinski in the corporate head office in Human Resources and Training. Later, she joined the education sector with the topmost universities IMS Unison Group and Amity University, the leading university in Pan Asia. She has obtained her master's in human resources & Management with Distinction. Prior to that, she graduated in economics with honors and law with a distinction, with an added vocational course in mass communication and analytic certification from IIM. Having an enriched research portfolio with Scopus and international refereed journals, book chapters, and conference presentations, she has been awarded as Best Presentation Award in international conferences like IEOM, MIDAS, and many more. She also serves as a member of AIMA and acts to lead and liaise between the student community and the industry delegates, with a keen interest in training and building sustainable models for business and society.

Dr. Richa Goel is an associate professor-economics and International Business at SCMS Noida, Symbiosis International Deemed University, Pune. She is a Gold Medalist in her master of economics with dual specialization at the master level accompanied by an MBA in HR and also with dual specialization at the graduate level in economics and Law. She has a Ph.D. in Management, where she has worked for almost 6 years in the area of Diversity Management. She has almost 20+ years of experience in academics along with authoring more than 50+ research papers for publication in reputed national and international journals accompanied by hundreds of papers for international/national conferences and symposiums. She has written more than 25 book chapters for leading publishers, and she has served as a member of the review committee for conferences and journals. She is also currently a series editor for two different series for Taylor and Francis.

Ms. Jan Alexa Sotto is a co-founder of Auspicia Inc., a manufacturer and seller of laboratory materials. She is a global businesswoman, a researcher, and an innovator in the field of innovation and IT. Having more than 10 years of experience in innovative business management, she has been actively involved in societal development in the Philippines in the field of medical aid.

1 Technology Transformation for Sustainable Research Pedagogy
An Indian Perspective

Madhuri Yadav and Pushpam Singh
Research Scholar, Department of Commerce, Indira Gandhi National Tribal University (A Central University), Amarkantak, Madhya Pradesh, India,

Sukanta Kumar Baral
Professor, Department of Commerce, Faculty of Commerce & Management, Indira Gandhi National Tribal University (A Central University), Amarkantak, Madhya Pradesh, India

1.1 INTRODUCTION

Today, education is a fundamental asset for achieving all types of potential of human beings related to economic, social, developmental, cultural, equality in social justice, and cyclic advancement of national education and cultural preservation in India. Therefore, the Central Government of India has given more importance to a unique National Education Policy-2020 (NEP-2020) and digital transformation of blended learning through a strategic plan on teaching-learning pedagogy for providing a high quality of educational opportunities to all from primary education to higher education. In a variety of contexts across the nation, high-quality, individualized learning approaches are proving to show significant improvements in student performance. Educators have access to a wide range of interactive learning resources and digital learning tools on which to base their models. A growing portion of the most impressive progress can be found in traditional school districts that are adopting models created around their own educational needs in accordance with the requirements therein, as the transformative benefits and outcomes these programmes can deliver are drawing more attention from the media, policymakers, and others.

Therefore, the aforesaid research title has encouraged us to acquire knowledge and skill in various segmentations, where expected outcomes will be possible

DOI: 10.1201/9781003365525-1

1

through meticulous planning and strategy, by which quickly and changing employment occurs in the country and world. This pedagogy of learning system must evolve to make education more experimental, holistic, integrated, innovative, learning centred, discussion based, and flexible in nature.

1.2   REVIEW OF LITERATURE

We concentrated on four key aspects of instructional design for blended learning in this section of the literature review: A strategic and ongoing assessment and evaluation of blended learning is required, as are (a) the thoughtful integration of face-to-face and fully online instructional components, (b) creative use of digital technology, (c) reconceptualization of the learning paradigm, and (d) a creative use of digital technology. The goal of the first principle is to make the most of the positive aspects of both environments and to better meet the requirements and preferences of the diverse student body [1]. Any technology should be used in a pedagogically suitable manner and for developing and maintaining socially located, highly participatory learning because of the inventive uses of technology [2]. The learning paradigm must be reconceived to include new pedagogies, learning theories (e.g., social constructivism and student-centred learning), the development of new understandings and knowledge through students' social interactions with a community of peers, and new roles for both students and teachers (e.g., mentors and coaches) [3]. In order to guarantee the quality of education, the fourth principle of persistent assessment and evaluation of blended learning solutions [4] is followed. In this chapter, we are attempting to demonstrate how NEP-2020 can benefit from a research pedagogy of blended learning and digital technology for the delivery of quality education and the enhancement of one special education policy in India for better employability prospects.

1.3   RESEARCH OBJECTIVES

We, the researchers, have taken the following objectives for enhancing the quality of blended learning and digital technology transformation in teaching pedagogy as related to NEP-2020:

- To study the role of research pedagogy of blended learning and digital technology for transformation of a learning system related with NEP-2020.
- To study NEP-2020 and its teaching-learning pedagogy.
- To study the research gap between the current state of learning outcomes and the requirement for future higher education in India.
- To find out the role of technological transformation in sustainable research and its challenges.

1.4   RESEARCH METHODOLOGY

In this section, we have followed the research tools and techniques as other researchers are followed previously such as Research Design, Method of Study,

Hypothesis Testing, Result Analysis, etc. But we have proposed both the methods of data collection as primary and secondary ways in order to get the information from the respondents through the Google form shared online with them, and we collected the response through a suitable questionnaire.

1.5 RESEARCH PROBLEM

The researchers have taken the research title "Technology Transformation for Sustainable Research Pedagogy" now associated with NEP-2020 (India) for providing quality education from Primary to PhD through the NEP-2020 policy in order to overcome the pitfalls in the education system at present and justify the research work.

1.5.1 FORMULATION OF THE PROBLEM

1.5.2 PROBLEM 1

To find out the pitfalls and causes that are preventing an integrated education policy in India and justify how an integrated and meticulous research pedagogy will execute NEP-2020 for providing quality education from primary to PhD.

1.5.3 PROBLEM 2

To execute a research pedagogy of blended learning and digital technology for transformation of learning through NEP on the basement of current teaching-learning pedagogy for a sustainable and qualitative teaching-learning process adopted by various universities in their curriculum.

1.6 HYPOTHESIS

In this section, the researcher has taken two hypotheses related to the aforesaid research design and title.

1.6.1 NULL HYPOTHESIS-1 (H_0)

NEP-2020 is not required for the present teaching-learning process in India due to diversity in language and culture.

1.6.2 ALTERNATIVE HYPOTHESIS-1 (H_1)

The teaching-learning pedagogy, as described by the Central Government related to NEP-2020, will be highly impactful for enhancing a better quality of the education system in India.

1.7   WHAT IS BLENDED LEARNING?

As blended learning is concerned, it is a formal educational programme in which students are participating through an online mode of delivery of instruction and content apart from their professor, guide, teacher, and mentor in order to control the student over time, Place, Path, and Pace are at least part of the guidance in the supervised brick-and-mortar dorm.

1.8   BASIC DIFFERENCE BETWEEN BLENDED LEARNING AND DIGITAL TECHNOLOGY LEARNING

The following are the basic differences between blended learning and digital technology.

Basic Difference	Blended Learning	Digital Technology
a. On Class Room based teaching	a. Blended learning is a combination of learning at a distance and the traditional on-campus learning (in a classroom). Basically, you will have a (more or less) fixed schedule where you will have to attend a part of the classes on campus.	a. Digital technologies are changing why, what, and how people learn.
b. On the basis of an Approach	b. Blended learning refers to an approach that combines e-learning with traditional in-person learning (think of lectures, workshops, and training sessions) and independent study.	b. It is a new and digitally enabled approach to learning design and delivery, including networked, mobile, and social learning. It is a learner-centred approach to learning and service design.
c. On the basis of Training	c. A blended learning programme can consist of monthly in-person training days, weekly assignments, and frequent – if not daily – peer-to-peer discussions on a digital facilitation platform.	c. It is more practical strategies and pathways for developing your professional practice.
d. On the basis of Hybrid Learning	d. Blended learning may be called "hybrid learning" because it studies 1) the environment(s) in which learning takes place and 2) the nature of the learning experience.	d. It will be applied to digital learning approaches to design and create premium digital learning experiences.

1.9   THE MEANING AND CONCEPT OF BLENDED LEARNING

With the development of digital technology, blended learning acquires new dimensions and combines the most advantageous aspects of traditional in-person instruction and online learning. It is challenging for educators to agree on a single definition of blended learning given the framework for blended learning that exists today, which is replete with many blends and models. Combining in-person education with computer-mediated training to promote interactive and reflective higher-order learning is the most typical description of blended learning.

1.10   ELEMENTS OF EDUCATION IN BLENDED LEARNING

According to the San Carlos, California-based Education Elements (http://www.edelements.com), many of the top-performing blended learning schools in traditional school districts, public charter schools, and private schools across the nation have been planned and implemented by Education Elements. They released a study on student outcomes for 5,000 elementary and secondary education students in 9 partner school districts in 2014. These are some of the consequences for students:

- There are 25% growths in math on NWEA Measures of Academic Performance (MAP) above national norms.
- There are 54% growths in reading on NWEA MAP above national norms.
- There are 87% of teachers who provide more differentiated instruction.
- Students in blended classrooms completed 1.5 years of course content in one year.
- An education elements partner district, students in blended classrooms met growth targets in reading (35% gain) and math (47% gain) above students in non-blended classrooms teaching.

1.11   WHO NEEDS TRANSFORMATION LEARNING?

The idea of modernizing education with technology is not new – far from it [5]. Many people thought that the introduction of slate blackboards to schools in the Great Plains and prairie states in the 1840s would revolutionize education [6]. In reality, the modifications to teaching methods brought about by slate blackboards still heavily influence many American classrooms today.

1.12   STRATEGIC BLENDING: A CONCEPTUAL FRAMEWORK

Yoon and Lim redefine blended learning as Strategic Blended Learning and Performance Solutions in the current context of maximizing organizational performance while engaging the learner. According to their explanation, this kind of blended learning is a deliberate blending of delivery media (mostly face-to-face and other technologies) to enhance learning/performance solutions that are generated from the objectives and requirements of a business. Yoon and Lim created a conceptual framework that takes into account five connected stages that strategically connect an organization's demands, performance solutions, and delivery methods (instructional and non-instructional).

1.13 THE FIVE PROCEDURAL PHASES

A. **Strategy and needs analysis** – These long-term business and human resource strategies, tasks, employee requirements, work systems, expenses, benefits, and the current technological infrastructure are all examined in this phase.

B. **Performance solutions** – This part will take into account both instructional and non-instructional methods of learning reinforcement. Techniques other than education could involve resources, resources, rewards systems, or institutional assistance. According to the organization's or institution's performance goals, component display theory, and learning theories suggest that a blended learning plan should balance the use of technology with face-to-face interaction.

C. **Delivery media** – The precise e-learning/digital technologies and face-to-face learning design methodologies are identified at this stage of the process. To choose the best combination of strategies, the authors refer to the e-learning structures identified by Driscoll and Rossett.

D. **Strategic blending** – At this stage, the effectiveness of the instruction, the budget, the frequency of need, and learner expectations are taken into account in relation to the performance objectives of the company.

E. **Evaluation and improvement** – In this stage, the strategic blended learning activity's inputs and outputs are assessed. Efficiency, effectiveness, cost, and eventual achievement of the performance outcomes would all be taken into consideration while evaluating the solution.

1.14 MODELS FOR THE BLENDED LEARNING MODEL

There are four types of blended learning models that are applied in the classroom setting:

a. **Rotation Model:** This methodology is comparable to the learning stations concept; the primary distinction is that in this model, students engage in both in-person and online instruction from a teacher. There is a predetermined schedule for students' virtual classes that are held in real life.

b. **Flex Model:** The foundation of the Flex Model is online learning; teachers serve only as facilitators and offer support as needed when students switch between learning activities on a flexible timetable. The teacher is still there for most of the lessons, and students still learn in a brick-and-mortar setting, but homework is now turned in online. This concept gives both teachers and students the flexibility they both need by giving them more say over how they spend their time.

c. **Self-blend Model:** The students are given flexibility over what and when they learn under this paradigm, but only in addition to their normal classes.

In addition to their typical in-class studies, students can take courses online, giving them more control over what they learn. With their own learning management system, schools can provide these courses, enabling students to advance in their online classes during downtime.

d. **Enriched Virtual Model:** In this arrangement, students must travel to the school, but all instruction takes place online in the school's computer lab. Although education is conducted remotely online, facility supervision and on-site student assistance are provided by trained non-instructional specialists. In this situation, technology isn't just limited to computers; an online lab model offers the chance to utilize the advantages of other cutting-edge technologies, such as 3D printers, VR headsets, and more, in a virtual setting.

1.15 PRACTICAL APPLICATION OF BLENDED LEARNING

Almost 85% of businesses use blended learning for the development and/or delivery of educational content, according to a 2013 study of "Blended Learning Best Practices" by The Learning Guild [7]. More than 76% of respondents said that blended learning was more effective than classroom instruction, and 73% said that blended learning had a higher value for and influence on learners than non-blended procedures. In their blended programme, more than 36% of respondents used six to ten different components. Classroom instruction, interactive web-based training, email communication, self-paced content, and threaded conversation were the top five elements. The Learning Guild (2013) used a blend of learning applications. According to survey participants, however, the top five challenges to implementing blended learning were a lack of funding, not picking the appropriate approach, a lack of support from senior management, the incapacity of developers and/or trainers, and a lack of technological infrastructure [8].

1.16 RESPONSE MODEL OF BLENDED LEARNING AND DIGITAL TECHNOLOGY-BASED LEARNING SYSTEM

The research has taken four groups of respondents – educational institution, university, company, and offices – 200 respondents, each from 50 number of male and female, and recorded their responses as per the responses apart from e-mail and through a Google form. In some cases, a telephonic questionnaire was also taken. The following responses have been recorded in Table 1.1.

1.17 HYPOTHESIS TESTING

In this section, the researcher has taken 200 respondents from four segmentations such as educational institutions, universities, companies, and offices, and asked which students and employees are in favour of blended learning and the digital mode of learning. Most of the respondents responded positively to the research pedagogy for blended learning and digital technology, which is so much more useful for providing quality education in secondary and higher education. Due to the response being high and positive, both the hypotheses are accepted.

TABLE 1.1

Response Model of Blended Learning

Year of Survey	Organization/Company	Nature of Learning	Percentage of Response (Positive)	Out Come
2017–2021	Educational Institution	Blended Learning and Digital Technology	76%	Highly effective
	University	Digital Technology/Blended Learning	73%	Highly effective
	Company	The Learning Guild of Digital Technology	85%	Highly effective
	Office	Digital Technology	65%	Effective

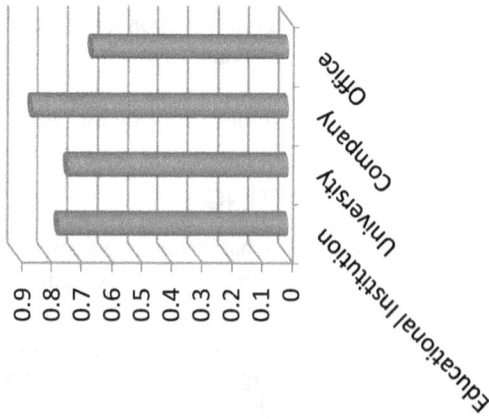

Percentages of four groups of Respondents in Pie Chart Table 1.1

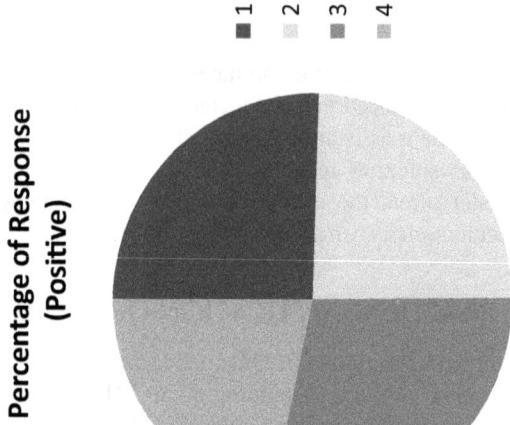

Responses of four groups in Barograph Table 1.1

Percentage of Response (Positive)

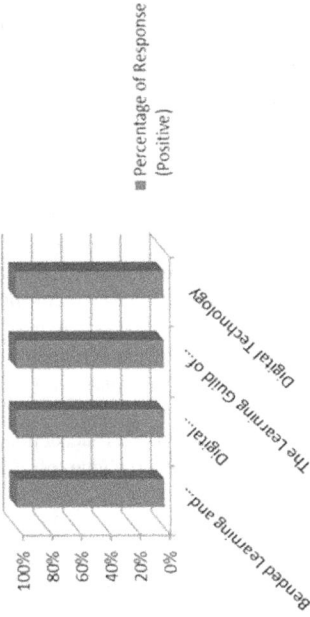

Semiotic Models of respondent for blended learning in pie chart

Percentage of Response (Positive)

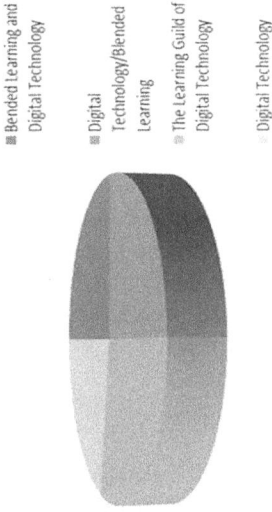

Semiotic Models of respondent for blended learning in bar graph

1.18   RESEARCH FINDINGS (OUTCOMES)

From the aforesaid research work, the researcher recommends the following:

- At present, the taken research pedagogy of blended learning has a great impact on the traditional mode of teaching and learning process.
- Digital technology and e-learning are also playing a vital role in imparting the virtual mode of education on an online and web-based learning platform that is enhancing the quality of research activity and learning process.
- NEP-2020 is an excellent learning pedagogy for providing an integrated learning system for primary to PhD, which enables nation building through the education system through both modes of the learning system.
- Digital or online learning has a better application while transformational learning is concerned, and it will be applied everywhere for providing quality education in both (offline/online) modes of teaching.

1.19   CONCLUSION

In conclusion, the above discussions prove that a research pedagogy of blended learning and digital technology for transformation of a learning system has a very positive effect in execution of NEP-2020 in India. There are simply too many unknowns in this landscape. In reality, because exceptional programmes view themselves as very early adopters, controversy has arisen about the designation of best practices in certain fields [9]. The majority of today's top educators uphold commitments to their instructional models' ongoing innovation and improvement. Such dedication to continuous improvement is necessary for the new pedagogy of individualized learning to reach its full potential [10, 11]. In businesses and educational institutions with established learning management systems, blended learning is a popular and successful way of workplace training.

REFERENCES

1. Al-Jarf, R. (2006). *Impact of Blended Learning on EFL College*, Riyadh: Readers, King Saud University.
2. Baker, C. (1991). *Foundations of Bilingual Education*. Clevedon: Multilingual Matters.
3. Barenfanger, O. (2005). Learning management: A new approach to structuring hybrid learning arrangements. *Electronic Journal of Foreign Language Teaching*, 14–35.
4. Cook, K., Owston, R. D., & Garrison, D. R. (2004). *Blended Learning Practices at COHERE Universities. (Institute for Research on Learning Technologies Technical Report No. 2004-5).* Toronto, ON: York University.
5. Das Neves Seesink, T. (2007). Using blended instruction to teach academic vocabulary collocation: A case study. Dissertation submitted to College of Education University of West Virginia.
6. Dziuban, C. D., Hartman, J. L., & Moskal, P. D. (2004). Blended learning. Research bulletin, 7. Retrieved April 27, 2008 from http://net.educause.edu/ir/library/pdf/erb0407.pdf
7. Cummins, J. (2001). *An Introductory Reader to the Writings of Jim Cummins* (C. Baker & N. Hornberger, eds). Clevedon: Multilingual Matters.

8. Lee, K. (2000). English teachers' barriers to the use of computer-assisted language learning, *The Internet TESL Journal*, Vol. VI, No.12.
9. Lightbown, P., & Spada, N. (1996). *How Languages Are Learned*. Oxford University Press, p. 135. ISBN 0-19437-168-7.
10. Heinze, A., & Procter, C. T. (2004). *Reflections on the Use of Blended Learning. Education in a Changing Environment*. Salford: University of Salford.
11. Graddol, D. (2007). *English Next – Why Global English May Mean the End of 'English as a Foreign Language'*. London: British Council.

2 Impact of Technology on Healthcare along with Social Security to Achieve Sustainability

Sheela Bijlwan, Manleenjot Kaur, and Ayushi Sharma
Assistant Professor, UIM, Uttaranchal University

Tilottama Singh
Associate Professor, Uttaranchal Institute of Management, Uttaranchal University

Neha Saini
Assistant Professor, School of Agriculture Uttaranchal University

2.1 INTRODUCTION

Employees' state insurance is a social insurance programme that protects workers in the organized sector (organizations with more than 10 employees) who make up to Rupees 21,000 a month. It fully covers their medical expenses (as well as unemployment benefits) in both government and corporate hospitals. As part of social security, health insurance for employees and their family must be provided by all employers in India by law. The implementation of internet-based technologies and artificial intelligence for better healthcare in India has been significantly impacted by the Era of Society 5.0, which offers opportunities and difficulties for healthcare service providers in each healthcare unit (Bogutska 2013). By enhancing functional effectiveness and creating and adopting innovative corporate models, services, and products, Industry 4.0 changed manufacturing and production systems. The goal of Society 5.0 was specifically to increase the productivity and sustainability of social systems. Further development is possible since the emphasis was placed on both the digitization and the automation of healthcare systems. The provision of healthcare faces a conflict between the requirement to maintain quality and the need to lower costs for the average person in emerging nations like India. Digital technology adoption can assist in finding a solution to these issues. According, to the study's findings, government effort ranks first, followed by

DOI: 10.1201/9781003365525-2

empowered customers and financial investment (Baker 2009). These elements produce a driving force so strong that it has a beneficial ripple effect. However, globalization and lifespan expansion, developing economies, worldwide rivalry, social and regional inequality, and other factors are further complicating the problem. The necessity of the hour is for sustainability across industries, green energy, climate control, and social innovation. The enormous potential of the Industrial Revolution 4.0 is laying the groundwork for the world to embrace Society 5.0, the desired future that can support as a bridge to a prosperous, digital-synchronized, Super Smart, human-centred society. Combining cutting-edge technologies like IoT, AI, robotics, Big Data, and advanced analytics, social innovation aims to create a successful society that balances social progress and economic growth. Initiating the development of Society 5.0, which converges cyberspace and real space by collecting Big Data from many sources via sensors and gadgets, persons, things, and systems are all interconnected. Big Data is examined by AI capabilities, which can coordinate back into the physical realm, with the innovative technologies through a variety of forms and media for individuals, businesses, and industries to simultaneously advance economic development and find answers to social issues (El Shafey 2020). In Society 5.0, the new value produced by social innovation eliminates regional, age, gender, and linguistic barriers and makes it possible to provide goods and services that are specifically tailored to each customer's needs and wants. It illustrates the capability to address several problems in numerous industries, including mobility, healthcare, agriculture, food, manufacturing, disaster management, and energy, among others.

2.1.1 Technological Innovation in the Healthcare Sector for Sustainability with Respect to Society 5.0

The Society 5.0 concept perfectly complements Hitachi's idea of a **"Sustainable Society,"** in which everyone can lead a secure and contented life. It is prepared to create a blueprint along with the support of the government for the gradual shift to Society 5.0 and assists in resolving many societal glitches through cutting-edge digital technologies. Hitachi has a strong and comprehensive portfolio, diversity of digital solutions, and integrated approach. As an important associate in the government's "Digital India" initiative, Hitachi is aiding India in acknowledging the demands of Society 5.0 in the coming era through mutual co-operation with the government in the forte of railways, finance, agriculture, urban development, and e-government.

2.2 TECHNOLOGIES THAT ARE BEST UTILIZED IN THE HEALTHCARE SECTOR

2.2.1 Intervention of IoT in the Healthcare Sector

The Internet of Things (IoT) offers an amazing arena to enhance communication between people and various physical and digital things, including areas of individualized healthcare. In Public health issues with chronic diseases brought on by risky infections like COVID-19 present the modern world with a few challenges. Due to the increase in health difficulties and high costs of healthcare, everyone is hopeful about adopting remote health management utilizing computer-aided technology, especially the old and disabled. The use of IoT potential, such as the Internet of Medical Things (IoMT) technology, which uses connected medical sensors or specialized medical equipment to give a personalized approach to healthcare delivery, can also fundamentally transform the environment of healthcare (Mourtzis et al., 2022). The use of IoMT technology in healthcare systems, also known as Healthcare Internet of Things (HIoT), brings through premium treatment opportunities for patients by joining the healthcare devices with the Internet and bringing forward services like the supervision of elderly people, telemonitoring, teleconsultations, and computer-assisted rehabilitation. However, what is more, crucial is the identification and prognosis of chronic disease at the initial stages. Systems for remote health monitoring made possible by the IoT have several benefits over conventional systems. Although it is crucial to accurately and efficiently collect raw data, what is more crucial is to explore and mine the data to uncover additional, priceless information, such as connections between goods and services.

Figure 2.1 depicts the role of IoT in the healthcare industry as it helps as follows:

1. It lowers the financial burdens on the patient and his or her family members.
2. It helps in maximizing better outcomes of treatments provided to the patients.
3. It also provides the best alternative methods of treatment.

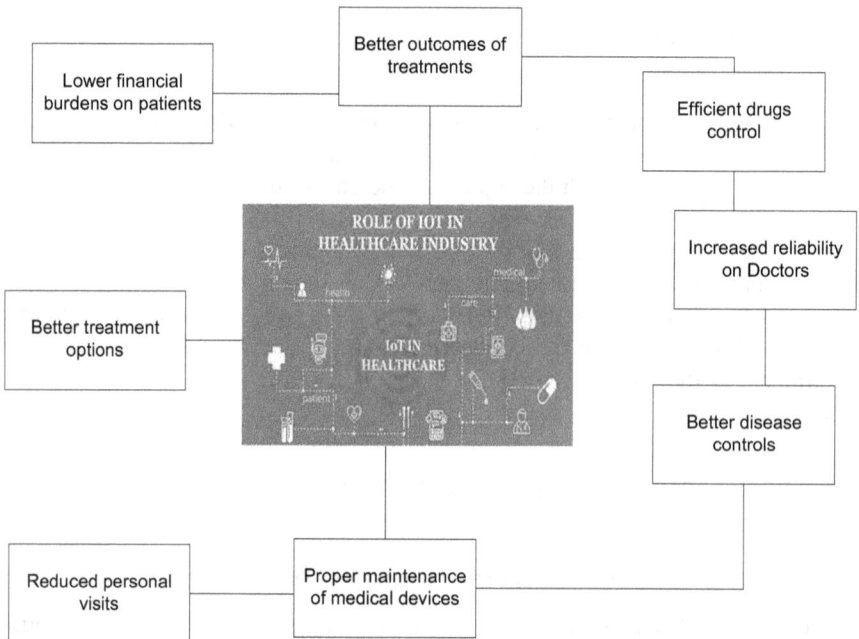

FIGURE 2.1   Role of Internet of Things in the healthcare industry.

4. IoT helps in reducing the personal visits of patients to the hospital.
5. It helps in proper maintenance of medical devices and builds strong trust in medical professionals.

2.2.1.1 Artificial Intelligence in Healthcare

AI and nanotechnology, when combined together, can act as the best innovation in the field of healthcare. It can act as early detection tools for any disease and assist independent living elderly people to live without fear. But, this advancement of technologies can face many difficulties in terms of moral values and privacy of individual or institutions. Thus, the sole responsibilities lie on the shoulders of policy makers and technology developers to be careful and responsible while implementing these technologies in any sectors (Figure 2.2).

Healthcare systems are in danger of collapsing unless substantial structural and transformative reforms are adopted. Larger staffing is also required for health systems; however, not only do more healthcare workers need to be educated and hired, but it's also vital to ensure that they devote their time to patients, where it is most valuable.

AI plays a vital role in medicine manufacturing process and remote patients' diagnosis. AI provides the strong basis for the smart health cities, which is the major outcome of SDG goals 3, to provide better healthcare services with sustainable development.

Nowadays, the technological intervention and integration of AI and IoT in the field of healthcare help in curing serious ailments like cancer and genetic disorders.

2.2.1.2 Machine Learning and Healthcare

AI is a division of machine learning (ML). Prototypes built on ML can learn through experimentation and natural learning without explicit programming. In other words,

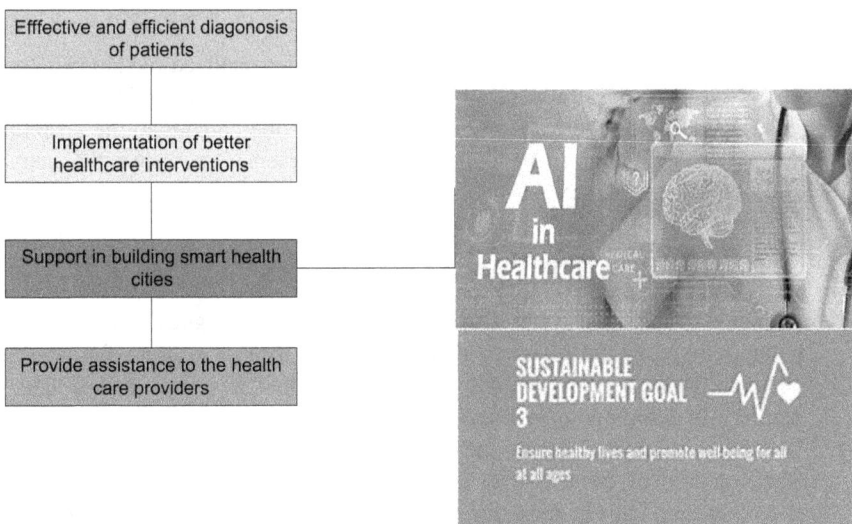

FIGURE 2.2 Depicts the application of AI in healthcare.

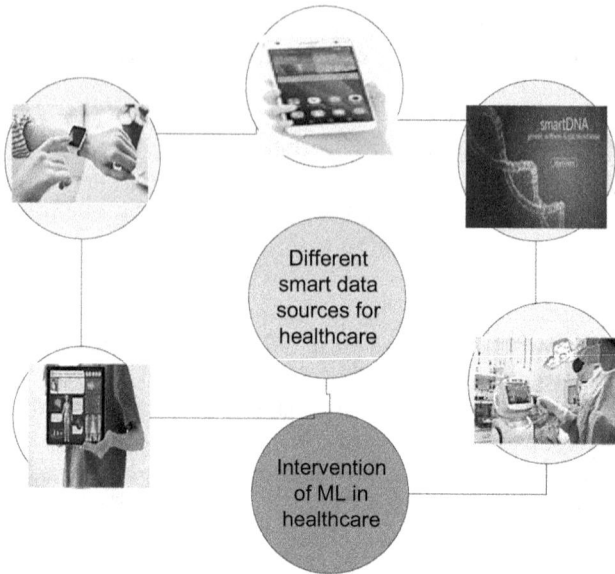

FIGURE 2.3 Intervention of ML in the health sector.

while explicit programming adheres to rules or a certain hypothesis, the learning model learns based on samples. ML increases computing process reliability and efficiency while lowering costs. Additionally, it can analyze data to develop models quickly and accurately. However, tools for processing massive amounts of data that are way above the capacity of human understanding are provided by ML. The resources used in healthcare for data creation are shown in Figure 2.3.

2.2.1.3 Application of ML in the Health Sector

ML is widely used in the health sector for providing personalized medicine as per the needs of the patients. It also provides electronic health records of the patients, which helps in reducing the frequent personal visits of the patients to the hospital. To elaborate its application, the diagrammatic representation shows the actual application of ML, in Figure 2.4.

2.3 RECOMMENDATIONS AND CHALLENGES

1. Utilization of AI and nanotechnology in the clinical diagnosis of critical patients helps in the effective management of disease without frequent personal visits to the hospitals. Moreover, virtual diagnosis and drugs delivery to the patients is one of the best advantages of AI, which the whole world realized during the recent COVID-19 pandemics.

2. The outcome of decades-long processes is the primary cause of death (heart disease, cancer, lung disease, and diabetes). These diseases can be prevented or delayed using current understanding. Lifestyle decisions that include a healthy diet, exercise, stress reduction, and vitamin supplements

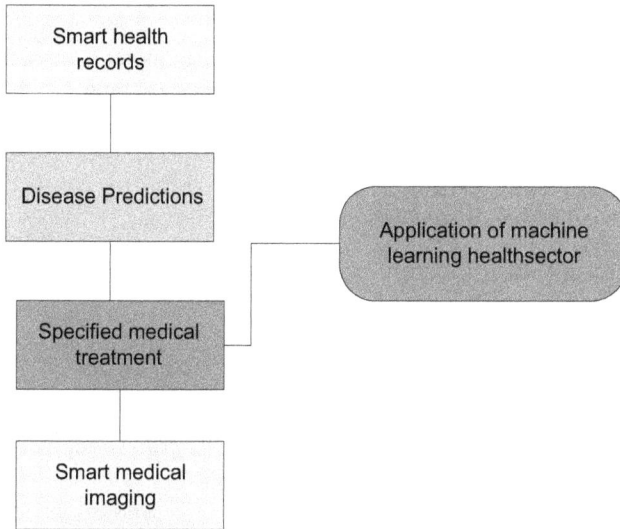

FIGURE 2.4   Depicts the application of machine learning in health sector.

can help with this. New genetic technology will enable people to create individualized plans, and cancer and heart disease early diagnosis will increase longevity. Moreover, the innovative green AI and green nanotechnology are totally based on environmentally friendly technology, which is highly required.

3. There is a future scope for the herbal medicines and ayurveda where AI and nanotechnology-based herbal medicines can be developed to treat diseases like cancer and chromosomal mutations in human beings and animals.

4. The key difficulty for AI/Nanotechnology in biomedicine and healthcare is not only creating unique, useful approaches but also making sure that they are accepted in daily clinical practice in the actual world. We anticipate that AI will soon become proficient at making diagnoses and therapy suggestions based on imaging.

5. This study primarily provides an overview of the application of AI and nanotechnology in the healthcare industry; however, we haven't gone into great details with other technologies like green AI and nanotechnology.

2.4   CONCLUSIONS

When it comes to real-world clinical trials, we are at a turning point in the evolution of traditional healthcare practice and technological application. When it comes to successful planning and treatment, higher accuracy and shorter turnaround times are essential for diagnostic care. However, integration of AI and nanotechnology has a vast and varied realm of data, analytics, neural networks, deep neural networks, and visualization methodologies that are constantly expanding and updating to meet the precise and accurate needs of the biomedical and health sectors. In addition, we need adequate funding and investment to educate healthcare professionals about the advantages and possibilities of the AI and nanotechnology-enabled healthcare system. The latest version of AI/ML and data science should be able to update extremely quickly and work with a small device in order to achieve high accuracy in customized, predictive, and portable healthcare systems. However, the implementation of these technologies must adhere to certain norms for data access, evaluation, validation, and ML, as well as for ethics, and privacy.

REFERENCES

1. Mourtzis, D., Angelopoulos, J., Panopoulos, N. A literature review of the challenges and opportunities of the transition from industry 4.0 to society 5.0. *Energies* 2022, 15.17.
2. Bogutska, K., Sklyarov, Y.P., Prylutskyy, Y.I. Zinc and zinc nanoparticles: Biological role and application in biomedicine. *Ukr. Bioorg. Acta* 2013, 1, 9–16.
3. Baker, J.R., Brent, B.W., Jr., Thomas, T.P. Nanotechnology in Clinical and Translational Research. In *Clinical and Translational Science*; Elsevier: Amsterdam, The Netherlands, 2009; pp. 123–135.
4. El Shafey, A.M. Green synthesis of metal and metal oxide nanoparticles from plant leaf extracts and their applications: A review. *Green Process. Synth.* 2020, 9, 304–339.

3 Smart Infrastructure

Role of Information Technology in Smart Cities

Seema Garg
Amity University, Noida, Uttar Pradesh

Ritu Sharma
Torrens University, Australia

3.1 INTRODUCTION

The last century saw a substantial improvement in the quality of life of individuals and society in general, particularly in terms of access to services. Technology has influenced this improvement many folds in the last few decades. Opportunities bring challenges, and thus administrators, architects, and urban planners have faced significant challenges as a result of the intense industrialization, introduction of technology, and rising population in metropolitan areas. Smart cities have become a future vision, which can be foreseen becoming a reality soon.

3.1.1 NEED FOR SMART CITIES

Smart cities are designed for optimum usage of space and resources along with an efficient and optimum distribution of benefits. They also aim to increase connectivity at various levels among citizens, as well as between the administration and population. They also originate in the 1980s prescriptions for managed, entrepreneurial cities – whose speed and flexibility in adapting to global markets make them more efficient. The model (Figure 3.1) of smart cities can be **represented as a community of average technology size, interconnected and sustainable, comfortable, attractive, and secure**. The landscape requirements and the solutions to local problems are the critical factors. In order to best serve their residents' present and future requirements, smart cities maximize the use of technology in the planning, construction, and maintenance of their structures and infrastructure. They must also take into account governance and development, urban planning and infrastructure, the environment and natural resources, society, and community in order to be really "smart."

The world's smartest cities were listed in the Smart City Index 2020, which ranked them according to economic and technological facts as well as the opinions of their residents.

DOI: 10.1201/9781003365525-3

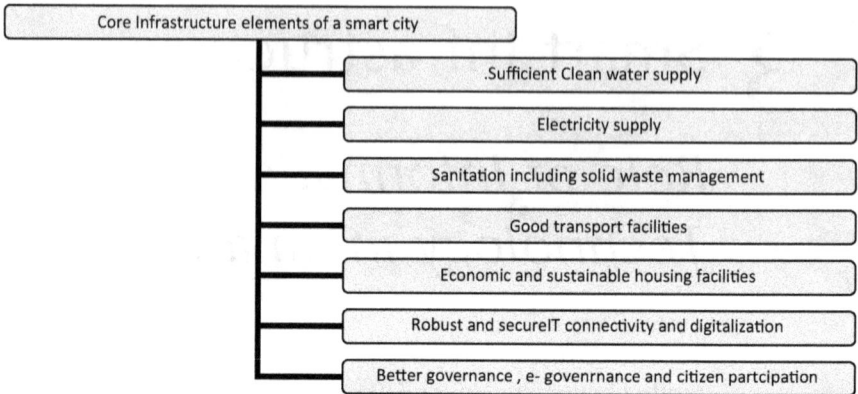

FIGURE 3.1   Core infrastructure of a smart city.

Singapore took the top spot in the ranking, demonstrating how a city can employ technology to address issues like population growth, climate change, and poverty. Helsinki, which is making efforts to lower its carbon dioxide (CO_2) emissions and enhance the quality of life for its citizens, came in second. The third position goes to Zurich, a city that has the highest standards for quality of life and is supported by an ecosystem of smart businesses that collaborate with many industries to introduce new projects.

Due to the increased concentration of people in cities, smart cities are quickly becoming a reality. The population of the world, which currently comprises 54% of cities, is projected to grow by 1.5 times by the year 2050. However, this growth also means a rise in energy use, which is bad for the environment; 60–70% of greenhouse gas emissions are caused by energy use in cities. Therefore, it is crucial to use energy efficiently in metropolitan settings. In addition, issues like increased traffic congestion, air pollution, crime, and environmental degradation are brought on by the concentration of people in urban metropolitan areas.

The priority for smart cities is the development of infrastructure, with the aim of improving the economy, the environment, social, cultural, and urban development. Therefore, cities should invest efforts in improving communication channels so that various services such as housing, entertainment, and telecommunications, among others, are interconnected.

3.2   STAGES OF THE IT INFRASTRUCTURE EVOLUTION

There are five key stages in the evolution of information technology infrastructure:

- The centralized mainframe.
- Personal computing.
- The client/server era.
- Enterprise computing.
- The cloud.

People are at the core of IT. People play a vital role in most operational systems and processes. Knowledge, skills, and attitudes often determine the quality and quantity of system output. ICT in the smart city is not only used to enhance the quality, performance, and interactivity of urban services, but also to reduce costs and resource consumption, and to improve contact between citizens and city stakeholders.

When a public service organization uses automation, artificial intelligence, cognitive computing, data sharing, analysis, and sensors effectively, it unquestionably lowers various operating costs, with mobilization costs being abruptly streamlined. Smart city components boost many aspects of the operation's performance, encouraging efficiency gains of up to 80% for certain procedures and an average of 25% gains throughout the entire physical networking. A speedy and confident management technique is made possible by a quicker and more informed decision-making process.

A quicker and more well-informed decision-making process enables a swift and confident management style.

By making more technology capable of communicating across platforms, Internet of Things (IoT) generates more data that can help improve aspects of daily life. Cities can identify both challenges and opportunities in real time, allocating resources more accurately to maximize impact by isolating issues prior to their emergence.

Cities are getting smarter. As a result, they are becoming more responsive and more livable – and today's world is merely at the cusp of what technology could eventually do in the urban environment. Indeed, across the world, we are seeing increased interconnectivity and people who are technology dependent. In response, a new wave of smart applications is changing how we approach everyday activities. Perhaps in your own home, you have personal assistants like Amazon's Alexa, an intelligent refrigerator, or a smart home security system. These types of technologies create opportunities for more efficient living. Smart cities use digital technologies and data to make better decisions, thereby improving the quality of life. These devices generate massive amounts of real-time data. More comprehensive, real-time data gives agencies the ability to watch events as they occur, understand how demand patterns are changing, and respond with lower-cost and faster solutions.

- The question of how data from sectors like air quality metres, public transportation, and energy generation can be combined and used effectively arises when residing in a smart city.
- IoT solutions might contribute to the solution. Forbes claims that by establishing a network of intelligent objects, a variety of technological advancements become possible, such as accurate traffic reports, real-time energy consumption data, and improvements to public transportation.
- IoT generates more data by enabling more technology to communicate across platforms, which can assist to enhance elements of daily life. Cities can recognize possibilities as well as obstacles in real time, allocating resources more effectively to maximize impact by identifying problems before they arise (Figure 3.2).

According to McKinsey, a smart city is created by combining three levels.

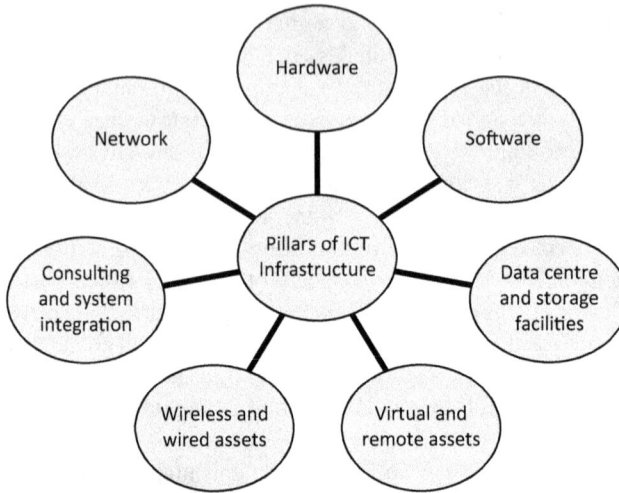

FIGURE 3.2 Pillars of IT infrastructure.

3.2.1 TECHNOLOGY

The technology base is the initial layer, and it consists of a significant number of sensors and smart devices connected by quick communication networks.

3.2.2 SPECIALIZED SOFTWARE PROGRAMMES

The second layer is made up of specialized software programs that convert unprocessed data into alerts, insights, and actions. These programs need the appropriate equipment, which is supplied by technology vendors and software programmers.

3.2.3 USAGE

Usage by businesses, cities, and the general public makes up the third layer. Many applications can only be effective if they are widely used and have the power to alter behaviour by motivating users to use public transportation outside of peak times, take alternate routes, use less energy and water (and do so at different times of the day), and practice preventative self-care.

3.3 AREAS OF IMPROVEMENT IN THE URBAN QUALITY OF LIFE

According to McKinsey's Global Institute Assessment about how smart city applications could affect various quality-of-life dimensions, including safety, time and convenience, health, environmental quality, social connectedness, and civic participation, jobs and the cost of living with some basic features are having the following effects:

3.3.1   IMPROVING PUBLIC SAFETY

- By utilizing a variety of technologies to their full capacity, it may be possible to reduce mortality from homicide, fires, and traffic accidents by 8–10%. Additionally, there may be a 30–40% reduction in the number of assault, burglary, auto theft, and robbery events. The advantages of these measurements are, of course, the comfort and mobility they would bring to city dwellers. Agencies can use data to deploy employees and limited resources more effectively when it comes to crime. For instance, statistical analysis is used to draw attention to patterns in real-time crime mapping. When crimes do happen, applications like home security systems, gunshot detection, and smart surveillance can hasten the response of law enforcement. Predictive policing can prevent crime from happening. However, the implementation of data-driven policing must avoid criminalizing particular neighbourhoods or demographic groups and must safeguard civil freedoms. Furthermore, smart technology may streamline field work and call centres when lives are on the line, and traffic-sign preemption can offer emergency vehicles a clear way. Applications of this kind could reduce emergency response times by 20–35%.

3.3.2   MAKING DAILY COMMUTES FASTER

- The daily commute is one of the key factors in raising quality of life. Cities that use smart transportation apps may reduce commute times by 15–20% on average by 2025. Of course, factors like the city's density, people's commute habits, and the state of the transit system all play a role in this. In general, programmes that streamline the experience for riders through digital signs or mobile apps that give real-time information about delays benefit cities with extensive, frequently used transportation systems. Additionally, placing IoT sensors on physical infrastructure already in place might assist technicians in resolving issues before they result in breakdowns and delays. In cities where driving and taking buses are the primary modes of transportation, apps that reduce traffic congestion are particularly useful. In developing cities with high bus usage, intelligent traffic signal syncing has the potential to cut average commute times by more than 5%. Drivers can choose the quickest route with the help of real-time navigation alerts about delays, and smart parking apps will direct them to open spaces.

3.3.3   BETTER PUBLIC HEALTH

- In the developed world, applications that support the monitoring, prevention, and treatment of chronic illnesses like diabetes or cardiovascular disease could have a significant impact. The health burden in high-income cities could be reduced by more than 4% thanks to remote patient monitoring technologies. These systems employ computerized gadgets to collect vital readings, which are subsequently transmitted to doctors for evaluation elsewhere. The patient and doctor can use this information to determine whether early intervention is necessary, which lowers problems and hospital stays. Cities can also use data and analytics to pinpoint interventions by identifying demographic groups with high-risk profiles. Health interventions can deliver messages about proper hygiene, risk-free sex, immunizations, and antiretroviral therapy adherence that could save lives. Data-based interventions that are concentrated on maternal and child health in low-income cities with high infant mortality rates could lower disability-adjusted life years (DALYs) by more than 5%. An additional 5% reduction is feasible if developing cities employ surveillance systems for infectious diseases to keep up with rapidly spreading outbreaks. Finally, telemedicine, which is becoming more popular in the wake of COVID-19, can be a life-saving strategy used in low-income areas with a doctor shortage.

3.3.4   CLEANER AND MORE SUSTAINABLE ENVIRONMENTS

- As urbanization, industry, and consumption rise, environmental pressures increase. Dynamic electricity pricing, some mobility apps, and building automation technologies might all work together to reduce emissions by 10–15%. Tracking water use combines sophisticated metering with digital feedback messages. In urban areas where residential water usage is high, it can motivate people to practice conservation and cut consumption by 15%. Water leakage from pipes is the main cause of water waste in the developing countries. It is possible to reduce these losses by up to 25% by using sensors and analytics. Applications like pay-as-you-throw digital tracking can cut the amount of solid waste generated per person by 10–20%. Cities can save 25–80 litres of water per person per day and reduce unrecycled solid waste by 30–130 kilos annually per person. Air quality sensors can locate sources of pollution and serve as the foundation for further action on the front of air quality health. Beijing achieved this reduction in harmful airborne pollutants by about 20% in less than a year by continuously monitoring the sources of pollution and adjusting traffic and development accordingly. Furthermore, people can take preventative measures when real-time air quality information is made available to the public via smartphone apps. This can lessen harmful health effects by 3–15%, depending on current pollution levels.

3.3.5   ENHANCE SOCIAL CONNECTEDNESS

- MGI conducted a study of urban inhabitants to ascertain the potential impact of digital platforms like Nextdoor and Meetup as well as digital methods for interacting with local government officials. According to McKinsey's estimate, these applications might almost triple the proportion of citizens who feel linked to local government and double that proportion to the local community. On the other hand, creating avenues for two-way contact between neighbourhood organizations and the general public might increase the responsiveness of local governments. Many municipal orga-nizations are engaged on social media, and some have created their own interactive citizen apps. These platforms provide platforms for residents to gather data, voice concerns, or provide input on planning issues in addition to disseminating information. For instance, Paris has implemented a par-ticipatory budget whereby anyone is invited to post project ideas, and then online votes are held to determine which ones need funding.

3.3.6 SMART CITIES ARE THE FUTURE

- Urban centres will grow, and smart technology will advance, resulting in their convergence. The infrastructure of cities can be improved by using Big Data and other intelligent technology to make better judgements. Of course, making better selections could have a big impact on the economy.
- As smart cities improve our quality of life, the potential cost savings and enhancements to various public service systems will only serve to increase demand for them. By focusing on the future, we can change how we interact with the environment as a whole and how cities interact with us, ensuring that we have access to the highest-quality options and use the fewest resources possible.
- Embracing digital means different things to different businesses, which is why Stefanini is here to assist you transform your particular firm.
- We provide a wide range of services to develop the ideal solution for your company. Industry 4.0, Cybersecurity, Cognitive, Customer Experience (CX), and other areas are among our specialties. We provide automation, artificial intelligence (AI), and user experience (UX) help to boost pro-ductivity and satisfy client requests.

3.4 ENGINEERING

It has traditionally been the responsibility of the civil engineer to take advances in science and technology and put them to use practically for the good of society. It has, out of necessity, been a deliberate, evidence-based strategy constrained by codes of practice and legislation. These conventional working methods are now being put to the test by the rapid advancements in information and communications technology (ICT), particularly in the usage of computer hardware and software.

It is essential to make sure that city infrastructure is adaptable and future-proof as urbanization rises and technologies that can enhance city efficiency and quality of life develop at an accelerated rate. This is made worse by the requirement to think about the effects of climate change and the requirement for both adaptability and resilience.

The real benefits of technology only come about as a result of high-quality engineering, not because technology is displacing the need for it. Engineers with international experience and knowledge of how various environments and cultures necessitate various solutions, the difficulties of new construction versus retrofit, and the demands of shifting regulatory and fiscal policy are better equipped in terms of skills, experience, and flexibility to deliver forward-thinking, future-proofed solutions for development and infrastructure projects.

Integral "layers of smartness" must be combined, including not only resource efficiency but also the promotion of health, economic stability, a sense of shared community, and the capacity to adapt to new challenges. This calls for a more sophisticated and general solution – an integrated strategy for urban development, finance, and governance in the language. The recent global economic downturn has exposed numerous developments that disregarded social and economic needs in the mistaken belief that any novel, audacious project with the guise of environmental sustainability would draw in customers.

3.4.1 SoS ENGINEERING

This chapter addresses the engineering of the IoT and how they interact to build a System of Systems, or SoS, in the complicated setting of smart cities. A use case involving smart streetlights is taken into account for this. As a result, this study elaborates on the building of a sophisticated SoS based on smart streetlight IoTs.

Model-based systems engineering (MBSE) is the SoS engineering methodology (Micouin, 2014). A workable engineering process is required in order to enable an organized approach to SoS engineering (Urgese et al., 2020; Delsing, 2017b). A Service Oriented Architecture (SOA) perspective can be used to describe this engineering process. Due to the use case architecture's foundation in SOA/microservices, it adheres to the main architectural trends in the field. An SOA framework and implementation platform are used in further exploration of a smart city engineering process. The engineering procedure is based on some fairly recent advancements in SysML-based SOA modelling (Delsing et al., 2021; MagicDraw). Thus, the Eclipse Arrowhead framework and implementation platform are used to further describe and discuss the engineering process (Cosgrave, 2017; Delsing et al., 2021).

3.5 DELIVERING BEST PRACTICES, PRODUCTIVITY, AND EFFICIENCY

The link between applied technology and infrastructure, city planning, and urban design must be made by engineers. They must give strong leadership within the construction sector and mediate the best solutions to competing challenges.

The following are the essential elements of this strategy: Effective communication to decision makers of integrated benefits from conception through operation and decommissioning; a clear understanding of how data function and are transferred between technologies and systems; a clear understanding of how data are captured and used by decision makers and the social and security implications of this; and a clear understanding of how technology can be used across functions to create economies of scope and scale.

Regarding the last point, future-proofing and value for money present major obstacles for potential clients, who must figure out how to guarantee that hardware installed today won't be outmoded in five years. Similar to the computer industry, real development and investment value are frequently best fostered through competition brought about by an open architecture/ecosystem-oriented approach, an approach driven by dynamic start-up businesses rather than the large, international corporations that tend to only push cities to adopt their solutions.

3.6 CYBERSECURITY CONSIDERATIONS IN SMART CITIES

Cybersecurity means protecting systems, networks, and programs against digital attacks. Cybersecurity protects computers, servers, mobile phones, and electronic systems from malicious attacks, and its function can be summarized: protecting the devices that people use, protecting the information on these devices, and protecting the identity of the people who use this information (Uchendu et al., 2021).

The smart city is a combination of smart devices, smart infrastructure, smart architecture, smart government, and the IoT. The interconnectivity of all these smart devices and the virtual and physical infrastructure that constitutes a smart city also gives rise to significant cybersecurity risks. Rarely in this rapidly developing field is security fully considered.

As a huge amount of sensitive data is exposed, despite taking all the security measures, there is a high risk of cyberattacks. Some of the attacks that can impair the working of smart cities are malware attack, phishing attacks, DDoS attacks, spoofing, ARP poisoning, rail network, traffic signal jamming, remote execution, replay attacks, eavesdropping, crypanalysis, tag killing, and ransomware attacks. It can also lead to data leakage, which can result in the misuse of the population's personally identifiable information.

A threat landscape of a smart city (Lévy-Bencheton et al., 2015) is shown in Figure 3.3.

A cyberattack in 2021 on the Florida water supply attempted to poison the water by briefly changing levels of lye in the drinking water, which was a critical intrusion into the infrastructure (Robles & Perlroth, 2021). Such cyberattacks on infrastructure can lead to a shutdown of crucial services and lead to significant damage.

Confidentiality can be endangered and ruptured by a large number of practices that are generally treated as intolerable; however, these are part of operations in a smart city ecosystem (Nautiyal et al., 2019).

The first step in overcoming cybersecurity problems in smart cities and protecting citizens' privacy is to identify cybersecurity challenges and threats to citizens' privacy. Without being aware of such challenges, and providing appropriate solutions, one cannot expect successful design, implementation, and development of a smart city (Kashef et al., 2021).

What can be done to help reduce cyberattack risk in a smart city (Das et al., 2019):

- Risk assessment: Threat modelling to identify risks.
- Risk assessment: Review risks for mitigation and acceptance.
- Continous risk assessment and management.
- City leader education to support the management and priorities.

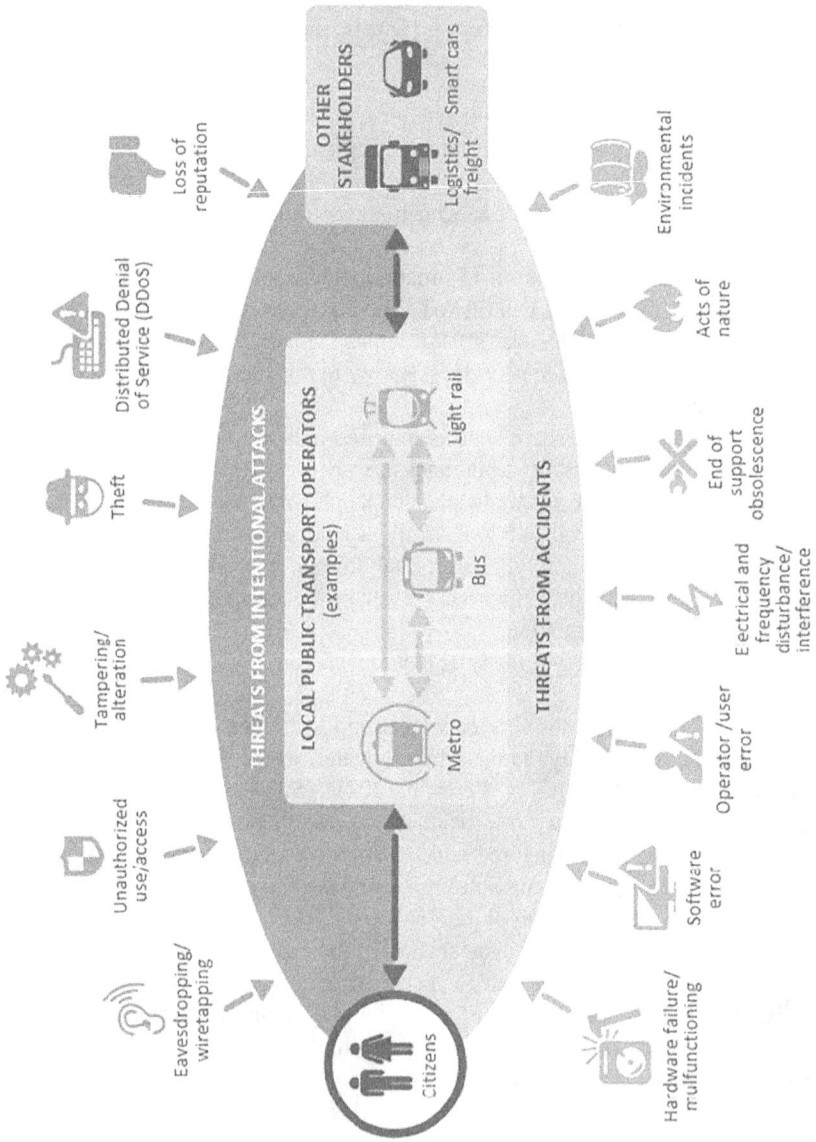

FIGURE 3.3 A threat landscape of a smart city (Lévy-Bencheton et al., 2015).

- Resilience consideration.
- Establishing minimum baselines for security.
- Training the workforce extensively and continuously.
- Educating citizens about cyber risks.
- Apply national as well as international standards for sharing the information.
- Run test drills to test established security plans.
- Create a plan in case of an attack for the risks that couldn't be mitigated by creating incident response capabilities and processes.
- Create a Computer Emergency Response team.

Cybersecurity framework for smart cities: (Waheed & Shafi, 2019)

To manage and improve cybersecurity, metaphysics-based security instruments are monitored, analyzed, and classified in this prescribed cybersecurity system.

The framework is comprised of three layers that deal with security.

Design time: It offers security through a service design and adaptation methodology.

Run time: This requires monitoring the IoT environment's network and processes for threats and weaknesses.

Integration layer: This allows reasoning methods for delivering appropriate security services that can be altered at design time, adapting to and acting with IoT environments. It enables both run time and design time to deal with the data or information about IoT security ontology (Duc et al., 2014) (Figure 3.4).

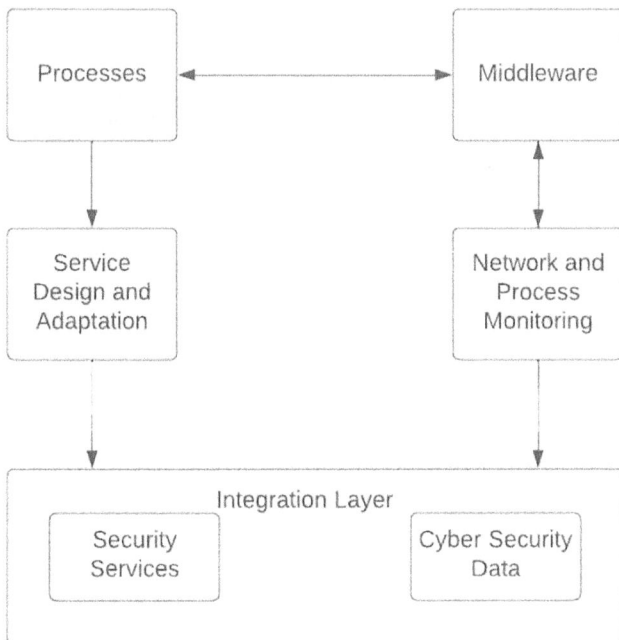

FIGURE 3.4 Cybersecurity framework.

3.7 CONCLUSION

The implementation of smart cities is complicated and not just reliant on technology. It must be supported by a carefully thought-out business case and incorporate integrated urban design concepts that take into account the specific local context. Collaboration between local and foreign consultants who are dedicated to extending the bounds of best practice is necessary. According to the most recent research on smart cities from IoT analytics, successful smart cities differ in seven ways. Inclusion of citizen empowerment, facilitation of public–private collaboration, and alignment with government initiatives are among the seven success factors.

There aren't many smart city implementations in existence today. International planners, designers, and engineers who have worked on developments around the world that are future-proofed and who use evidence-based approaches to urban design must provide developers and municipalities with enough assurance that the concept is feasible. Smart cities have come to light as a potential remedy for sustainability issues brought on by rapid urbanization. They are thought to be necessary for a sustainable future. Also, the Smart Green Cities is a collaborative hub connecting industry, government, researchers, and community to create liveable urban environments by inspiring change through evidence-based problem solving. Cities are dynamic and complex by their very nature. As a result of providing a secure environment for the infusion of capital, smart cities are appealing to investors and entrepreneurs in addition to their residents.

REFERENCES

Caragliu, A.; Del Bo, C.; Nijkamp, P. Smart Cities in Europe. *J. Urban Technol.* 2011, *18*(2), 65–82.

Catriona, M.; Cochrane, G.; Cave, J.; Millard, J.; Pederson, J.K.; Kåre, R.; Liebe, A.; Wissner, M.; Massnik, R.; Kotternik, B. Mapping Smart Cities in the EU: Directorate-General for Internal Policies. Policy Department A: Economic and Scientific Policy. European Parlament 2014.

Colombo, A.W.; Bangemann, T.; Karnouskos, S.; Delsing, J.; Stluka, P.; Harrison, R.; Jammes, F.; Lastra, J.L. Industrial Cloud-Based Cyber-Physical Systems. *IMC-AESOP Approach* 2014, *22*, 4–5.

Cosgrave, E. The Smart City: Challenges for the Civil Engineering Sector. *Proc. Inst. Civ. Eng. Smart Infrastruct. Constr.* 2017, *170*, 90–98.

Das, A.; Sharma, S. C. M.; Ratha, B. K. (2019). The New Era of Smart Cities, From the Perspective of the Internet of Things. In *Smart Cities Cybersecurity and Privacy* (pp. 1–9). Elsevier.

Delsing, J. (Ed.) *IoT Automation-Arrowhead Framework*; CRC Press: Boca Raton, FL, USA, 2017a; ISBN 9781498756754.

Delsing, J. Local Cloud Internet of Things Automation: Technology and Business Model Features of Distributed Internet of Things Automation Solutions. *IEEE Ind. Electron. Mag.* 2017b, *11*, 8–21.

Delsing, J.; Eliasson, J.; van Deventer, J.; Derhamy, H.; Varga, P. Enabling IoT automation using local clouds. In Proceedings of the 2016 IEEE 3rd World Forum on Internet of Things (WF-IoT), Reston, VA, USA, 12–14 December 2016; pp. 502–507.

Delsing, J.; Kulcsár, G.; Haugen, O. SysML Modeling of Service-Oriented System-of-Systems. *Innov. Syst. Softw. Eng.* 2021. submitted for publication.

Duc, A. N.; Jabangwe, R.; Paul, P.; Abrahamsson, P. (2017, May). Security challenges in IoT development: a software engineering perspective. In Proceedings of the XP2017 scientific workshops. (pp. 1-5).

Eclipse Papyrus-Modeling Environment, Eclispe Foundation. Available online: https://www.eclipse.org/papyrus/ (accessed on 29 April 2021).

ISO 811346-12:2018; Standard Document; ISO: Geneva, Switzerland, 2018.

ISO/IEC/IEEE 15288:2015; Standard Document; ISO: Geneva, Switzerland, 2015.

Jamshidi, M. System of Systems Engineering-New Challenges for the 21st Century. *IEEE Aerosp. Electron. Syst. Mag.* 2008, *23*, 4–19.

Karnouskos, S.; Colombo, A.W. Architecting the next generation of service-based SCADA/DCS system of systems. In Proceedings of the IECON 2011, Melbourne, VIC, Australia, 7–10 November 2011; p. 6.

Kashef, M.; Visvizi, A.; Troisi, O. Smart city as a smart service system: Human-computer interaction and smart city surveillance systems. *Computers Human Behav.* 2021, *124*, 106923.

Lévy-Bencheton, C., Darra, E., Bachlechner, D., Friedewald, M., Mitchener-Nissen, T., Lagazio, M., & Kung, A. (2015). Cyber Security for Smart Cities - an Architecture Model for Public Transport.pdf.

Lin, S.W.; Crawford, M.; Mellor, S. The Industrial Internet of Things Volume G1: Reference Architecture. Technical Report, Industrial Internet Consortium. 2016. Available online: http://www.iiconsortium.org/IIC_PUB_G1_V1.80_2017-01-31.pdf (accessed on 29 April 2021).

Lombardi, P.; Giordano, S.; Farouh, H.; Yousef, W. Modelling the Smart City Performance. *Innov. Eur. J. Soc. Sci. Res.* 2012, *25*(2), 137–149.

MagicDraw, SysML Modelling Tool, Dassault Systems. Available online: https://www.nomagic.com (accessed on 29 April 2021).

Maier, M.W. Architecting Principles for Systems-of-Systems. *Syst. Eng. J. Int. Counc. Syst. Eng.* 1998, *1*, 267–284.

Mennenga, M.; Cerdas, F.; Thiede, S.; Herrmann, C. Exploring the Opportunities of System of Systems Engineering to Complement Sustainable Manufacturing and Life Cycle Engineering. *Procedia CIRP* 2019, *80*, 637–642.

Micouin, P. *Model Based Systems Engineering: Fundamentals and Methods*; John Wiley & Sons: Hoboken, NJ, USA, 2014.

Nautiyal, P., Rajput, R., Pandey, D., Arunachalam, K., & Arunachalam, A. Role of glomalin in soil carbon storage and its variation across land uses in temperate Himalayan regime. *Biocatal Agric. Biotechnol.* 2019, *21*, 101311.

Paniagua, C.; Delsing, J. Interoperability mismatch challenges in heterogeneous SOA-based systems. In Proceedings of the IEEE International Conference on Industrial Technology (ICIT), Melbourne, VIC, Australia, 13–15 February 2019.

Standards IEEE 1547, IEEE 2030; Standard Document; IEEE SCC21; IEEE: Piscataway, NJ, USA, 2021.

Tizzei, L.P.; Azevedo, L.; Soares, E.; Thiago, R.; Costa, R. On the maintenance of a scientific application based on microservices: An experience report. In Proceedings of the 2020 IEEE International Conference on Web Services (ICWS), Beijing, China, 19–23 October 2020; pp. 102–109.

Urgese, G.; Azzoni, P.; van Deventer, J.; Delsing, J.; Macii, E. An engineering process model for managing a digitalised life-cycle of products in the Industry 4.0. In Proceedings of the NOMS 2020—2020 IEEE/IFIP Network Operations and Management Symposium, Budapest, Hungary, 20–24 April 2020; pp. 1–6.

Waheed, A., & Shafi, J. (2019). Efficient cyber security framework for smart cities. In *Secure Cyber-Physical Systems for Smart Cities,Advances in Computer and Electrical Engineering* (pp. 130–157). IGI Global.

Wenge, R.; Zhang, X.; Dave, C.; Chao, L.; Hao, S. Smart City Architecture: A Technology Guide for Implementation and Design Challenges. *China Commun.* 2014, *11*, 56–69.

4 Chatbots Communication Quality Impact on Brand Experience of Customers
An Interdisciplinary Study

Priyanka Tyagi
Research Scholar, Faculty of Management,
SRM Institute of Science and Technology,
Delhi NCR Campus, Uttar Pradesh, India

Baljit Kaur
Assistant Professor, Faculty of Management,
SRM Institute of Science and Technology,
Delhi NCR Campus, Uttar Pradesh, India

Narendra Mohan Mishra
Professor, Faculty of Management, SRM Institute of Science
and Technology, Delhi NCR Campus, Uttar Pradesh, India

4.1 INTRODUCTION

Artificial intelligence (AI) is generally referred to as "augmented human intelligence" (Romero-Brufau et al., 2020). One of the current and emerging trends in AI is the use of "chatbots," which are software programs that converse with users. The use of chatbots has grown in prominence recently as a means of customer interaction (Johannsen et al., 2021). Online interactions are now a popular way for businesses and customers to communicate (Steinhoff et al., 2019).

The chatbot agents assist organizations in enhancing customer experiences and exceeding expectations by interacting with customers (Yen & Chiang, 2021). The gap in the literature is filled through this present research. The present study determines the impact of credibility, openness, and capability of chatbot communication on the brand experience of customers. A number of the studies ignored this aspect. Sung et al. (2021) observed that AI technology increases customer engagement and, also has a favourable effect on intentions to share brand experiences with social groups.

DOI: 10.1201/9781003365525-4

Interactive AI technology opens up new potential to improve customer interaction. Følstad and Brandtzaeg (2020) opined that the users must find chatbots to be helpful and enjoyable. The current study is important because live chat interfaces are becoming an increasingly common way to offer real-time customer care on many e-commerce platforms. Conversational bots now frequently assume the responsibilities of human chat support agents. Businesses are using chatbots more frequently for customer assistance (Fotheringham & Wiles, 2022).

Although there is a lot of literature on AI, there is little knowledge of its connection to brand experiences. There is limited research on chatbot communication quality (credibility, openness, capability) and how they affect the brand experiences of customers. The researcher used theoretical information from the multidisciplinary literature on chatbots to fill in the aforementioned study gaps and advance the marketing field. This study aims to ascertain whether a chatbot agent's communication quality has any effect on the brand experiences of customers. The effect of chatbots on brand experiences could be positive, negative, or neutral. The organizations will benefit from this study by being able to evaluate the impact of the communication quality of chatbots on the brand experiences of customers. The remaining part of the chapter is structured as follows: Researchers start with a brief explanation of the literature that has already been published on the significance of chatbot agents, the gaps in the literature, the methodology used, and the in-depth analysis followed by conclusion and limitations.

4.2 LITERATURE REVIEW

AI is evolving rapidly, and this offers interesting potential for academic and practical marketing research (Mustak et al., 2021). AI may imitate people and carry out tasks in an "intelligent" way, according to an expanding body of research on AI in marketing (Vlačić et al., 2021).

The extensive utilization of information and communication technology, such as AI, Internet of Things (IoT), and blockchain technology, is accelerating the global technological and Industrial Revolution. Government, business, and academia have all given AI a lot of attention. AI requires not only rational reasoning and imitation but also an essential component of emotion. The next development in AI technology may enable computers to express emotion as well as increase their capacity for logical reasoning. Shortly, machine intelligence may surpass human intelligence (Zhang & Lu, 2021). The chatbots have evolved into useful tools for providing automatic, logical, and accurate responses to customers' inquiries. It has been noted that chatbots have offered users all over the world a variety of services, such as placing orders for food, recommending goods, giving insurance policy advice, providing customer support, providing financial assistance, scheduling meetings, etc. Marketing managers and researchers increasingly place a high emphasis on managing the brand experience of customers through technology (Witell et al., 2020). The use of AI in marketing is currently on the rise as a result of rising computing power, falling computing costs, the accessibility of massive data, and the development of machine learning techniques and models (Huang & Rust, n.d.-b). In recent years, a wide range of marketing researchers has shown considerable interest in AI (Kopalle et al., 2022). Both practitioners and academics agree that AI has the potential to offer considerable advantages (Grewal et al., 2021). AI is changing how businesses and customers interact. However, contemporary AI marketing representatives are frequently viewed as frigid and heartless and can be inadequate alternatives to human-based engagement (Liu-Thompkins et al., 2022).

Since the widespread use of voice assistants powered by AI in our daily lives, voice shopping is growing in popularity among consumers (Bawack et al., 2021). But companies do not even fully utilize AI's capabilities and most common techniques. This is particularly the case with marketing, as its potential application goes beyond simple decision making, customization, and segmentation (Volkmar et al., 2022).

There are a large number of B2B companies that have used chatbots powered by AI to offer human-like service contact at various customer touchpoints. The continuous provision of an improved, live-channelling-channel customer experience is one of the goals behind the implementation of this technology. A wide range of industries, including customer service, health, education, and marketing are rapidly adopting chatbots as key entry points for digital services and information (Oh et al., 2021).

The retail market is also being disrupted by AI. As one of the most widely used AI technologies, digital voice assistants have the potential to completely transform how consumers purchase, but our knowledge of how they may improve the brand experience is still limited (Aw et al., 2022). In the future of retail, smart technologies have the potential to significantly improve the brand experience. The retailers are increasingly using technology-enabled personalization techniques to improve the in-store customer experience (Riegger et al., 2022). AI has advanced the retail industry in numerous ways in recent years, including by making Big Data available for prediction, supporting more educated purchasing and consumption decisions, enabling visual goods displays, and fostering consumer engagement (Huang & Rust, n.d.-a). There is a huge amount of scope for experimenting with various methods employing AI-powered digital assistants to create a genuinely enjoyable social environment in-store. The retailers who engage in new AI in-store technologies might notice that it contributes to small but steady improvements in customer experience and a more noticeable impact of technology on customer experience (Moore et al., 2022). The customer experience will determine whether chatbots are used for customer service and marketing. The user experience for these chatbots is particularly concerned with whether the user receives appropriate responses to their questions and whether their engagement with the chatbot advances their quest for a solution (Følstad & Taylor, 2021).

Therefore, businesses need to reconsider their branding strategies and incorporate technology, especially those selling goods in low-involvement industries. They must take into account how rivals choose a voice tactic instead of a visual brick-and-mortar one (Klaus & Zaichkowsky, 2022).

Brand experience is a crucial component of achieving a competitive advantage in marketing management, particularly for service-based and B2B businesses (Kushwaha et al., 2021). Brands are looking for innovative and meaningful ways to engage consumers (Pallant et al., 2022). The objective of this present study is to determine the impact of the credibility, capability, and openness of chatbots on brand experiences. The researcher conducted a thoughtful study to attain these objectives.

4.3 RELATED LITERATURE AND HYPOTHESIS DEVELOPMENT

4.3.1 CREDIBILITY AND BRAND EXPERIENCES

The credibility of communication sources affects consumers' purchase decisions (Qi & Kuik, 2022). Given that all decisions are based on information, the value of information credibility in society cannot be ignored (Alrubaian et al., 2019). Chatbots received increased power to serve as a communication tool that disseminates information and upholds customer experience. The findings of the study demonstrated that knowledge of communication credibility is important in determining customer satisfaction and brand experiences. As a result, communication credibility is crucial for promoting customer satisfaction and brand experiences (Jansom et al., 2022). Further, businesses can get a competitive edge and achieve market success by having a strong brand image that encourages customers to pay higher prices. Oral communication is less effective in creating a strong brand image than online reviews, blogs, and texts on brand experiences. Hence, it is observed that credibility of communication impacts brand experiences of customers (Chakraborty & Bhat, 2018).

H_1: Credibility positively affects brand experiences of customers.

4.3.2 OPENNESS AND BRAND EXPERIENCES

The informal coordination in the form of communication between the human and the electronic brain has made significant progress. For the development of logical language and terminologies, electronic brain protection systems are frequently used. A chatbot functions as an electronic brain that enables human interaction by using logical language (Yadav & Dhanda, 2023). Further, by using open, affective, and coherent communication, chatbots have proved to encourage students (Huang et al., 2022). The number of consumer–chatbot service interactions has been growing, and chatbots are taking over more and more of the online retailing market (Xu et al., 2022). Interestingly (Meichan & Rui Wang, 2023), when the chatbots adopt an informal language in comparison to formal language style, customers' continuance usage intention and brand attitude increase.

H_2: Openness of chatbots has a positive impact on brand experiences.

4.3.3 CAPABILITY AND BRAND EXPERIENCES

An artificial agent and humans can communicate with each other through chatbots, which are computer programmes (Alsubayhay et al., 2022). A conversational agent known as a chatbot uses text or voice communications to replicate human communication (Khadija et al., 2021). Similar to a digital assistant, a chatbot can be programmed or can learn on its own. It may be taught to comprehend user inquiries and react in the user's native language during a discussion. Due to its capability to use natural language processing, chatbots can actively assist humans in participating in online conversations (Balaji & Yuvaraj, 2019).

In the future, chatbots' increasing capabilities will become increasingly essential (Camilleri & M.A, 2022).

H₃: Capability positively influences brand experiences.

4.3.4   RESEARCH GAP

Marketers have used chatbots to encourage relationship building as firms increasingly include AI-enabled automation as part of their communication efforts. Further research is necessary to determine how particular aspects of communication quality influence the brand experience of customers. There is a need to understand the impact of communication quality dimensions, specifically credibility, openness, and capability of AI technology-powered chatbots on the brand experiences, but no research has attempted to conceptually and empirically investigate the phenomena. There are no studies, in particular, that link the communication quality of chatbots to brand experiences. Because of this, it's critical to comprehend what influences customers' brand experiences. Research on AI-powered chatbots in marketing is essential to advancing the state of AI technology's current applications.

4.3.5   METHODS

The current study aims to investigate the effect of communication quality (credibility, openness, capability) of chatbot agents on brand experiences and inclinations to use their recommendations. A thoughtful study is designed to accomplish these goals.

4.3.6   SAMPLING FRAMEWORK AND QUESTIONNAIRE DESIGN

The sample population for this study consists of Indians, specifically Delhi NCR area customers. To add variety to the data, the respondents were from various demographic profiles. The participants were between the ages of 18 and 51 above, had a good mix of genders, and represented a wide range of backgrounds and qualifications in terms of education and employment. In the present study, 221 respondents participated in an electronic survey that collected the data for this investigation. Two sections make up the survey. The first section of the questionnaire included questions about the respondent's demographic profile, while the second section asked questions to determine how independent variables impacted the dependent variable. In the study, brand experience is the dependent variable, whereas communication quality (credibility, openness, capability) is the independent variable (Figure 4.1).

4.3.7   MEASURES

The measurement of constructs by measurement scale is a crucial aspect of research (Ringle & Sarstedt, 2016). Based on validity and reliability, a measurement scale's usefulness can be assessed. Based on previous literature, researchers developed 15 items to measure the communication quality of chatbots (Haugeland et al., 2022; Baabdullah et al., 2022). To measure brand experience, researchers developed 11 items after extensive literature review (Schmitt et al., 2008).

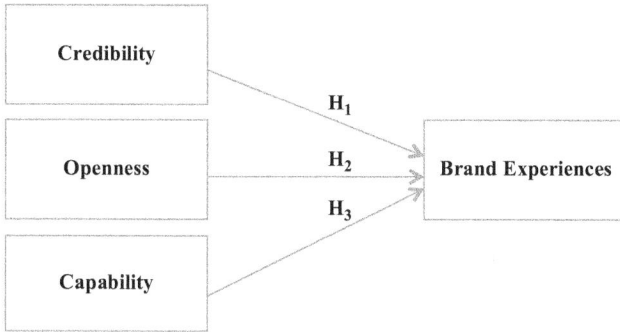

FIGURE 4.1   Research framework developed by researchers.

Researchers measured these items using a 5-point Likert scale (1 = "strongly disagree," 5 = "strongly agree").

4.3.8   RESEARCH TOOLS AND TECHNIQUES

The construct reliability was measured by using Fornell and Larcker criterion, Furthermore, CR value should be 0.7 or more, which shows an acceptable level of reliability. Additionally, AVE was used in the study to assess the convergent validity; AVE (average variance extracted) value should be above 0.50. The cross-loading values of the latent variables were used to obtain discriminant validity. The cross-loading values in corresponding variables should be higher in comparison to the loadings of other constructs. Further, to assess the suitability of the scale items, a measurement model has been employed. Moreover, to gauge the model's prediction accuracy, the coefficient of determination, or R^2, was used by the researchers (Hair et al., 2019).

4.3.9   ANALYSIS OF THE MEASUREMENT MODEL

The overall results for the measurement model were acceptable, with an NFI = 0.86 and SRMR = 0.03. In order to assess the convergent and discriminant validity of the observed variables, the researchers examined the measurement model. Accordingly, loading values, AVE values, and CR values must all be above 0.6, 0.5, and 0.7, as prescribed by the researchers (Hair et al., 2019). The fulfilment of these requirements assures a model's fitness. In Table 4.1, the measuring features of the constructs are described in detail.

4.3.10   ANALYSIS OF STRUCTURAL MODEL

The results of the structural model show the model's goodness of fit as well as predictive validity. The results show that perceived capability has a positive impact on the brand experience of customers (β = .745, p < .001), as well as openness (β = 0.174, p < .001), whereas credibility has less influence on the brand experience of customers (β = 0.075, p < .001). Since NFI = 0.861 and R^2 = 0.956. Thus, H_1, H_2, and H_3 were supported (Figure 4.2).

TABLE 4.1

Constructs, Factor Loadings, Cronbach's Alpha, Composite Reliability, Average Variance Extracted

Constructs	Factor Loadings	Cronbach's Alpha	CR	AVE
Credibility				
I believe the chatbot agent replies to me sincerely.	–	–	–	–
I believe the chatbot agent replies me reliably.	0.889			
I believe the chatbot agent treats me fairly and justly.	0.959			
I believe chatbot agents are trying to make a long-term commitment.	0.921			
I believe whenever chabot agents make an important decision.	0.925	0.978	0.978	0.862
I know it will be concerned about me.				
I feel the chatbot agents reply to me timely.	0.947			
I believe the chatbot agent replies to me accurately.	0.933			
I believe the chatbot agent answers me completely.	0.924			
Openness				
I can easily have a conversation with chatbot agents.	0.924			
I can easily express what I wanted to express with chatbot agents.	0.904	0.954	0.954	0.837
I can understand the answers from chatbot agents.	0.932			
I believe the chatbot agent attitude is friendly.	0.900			
Capability				
I believe this kind of chatbot agent can deal with complicated problems more efficiently than offline stores.	0.947			
I believe these kinds of chatbot agents can deal with complex problems efficiently.	0.933			
I believe this kind of chatbot agent usually saved my decision-making time.	0.914	0.972	0.972	0.874
I think the chatbot agent can solve the problem very well.	0.932			
I believe chatbot agents make the good purchase decision based on my interactions.	0. 947			
Brand Experience				
When I come across the brand on the advice of the chatbot agent, I do a lot of thinking.	0.942			
The brand communicated by chatbot agents induces my feelings and sentiments.	0.930			
The brands generally communicated by chatbot appeal to me in a sensory way.	0.874	0.966	0.966	0.824

TABLE 4.1 (Continued)

Constructs, Factor Loadings, Cronbach's Alpha, Composite Reliability, Average Variance Extracted

Constructs	Factor Loadings	Cronbach's Alpha	CR	AVE
My visual sense and other senses are greatly affected by the brand that chatbot agents represent.	0.889			
The brand usually communicated by chatbot agents entices my interest and helps me think critically.	0.922			
The brand usually communicated by chatbot agents is an emotionally engaging brand.	0.887			

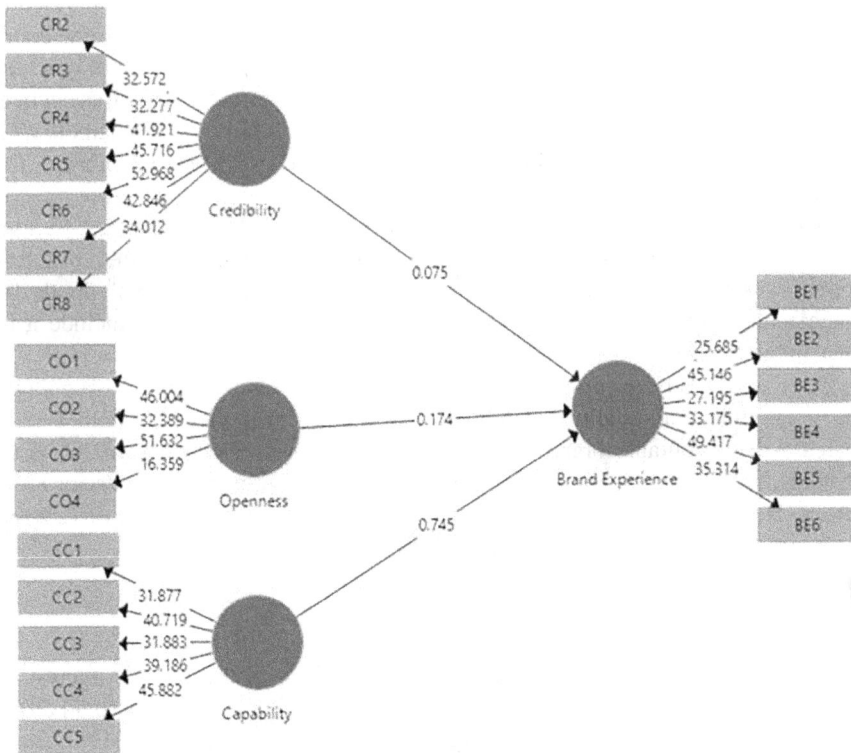

FIGURE 4.2 Results structural model.

4.4 CONCLUSION AND IMPLICATIONS

Businesses increasingly use chatbots as virtual customer service representatives on their websites and mobile applications but little academic research has been done on the branding consequences of such practices. A model was developed to assess their relationships realistically. The findings of this study show that the communication quality (credibility, openness, capability) of AI-based chatbots positively influences the brand experiences of customers. The gap in knowledge regarding the role of communication quality of AI-based chatbots in the context of brand experiences, an essentially complex and socially integrated activity, served as the driving force for this work. This study contributes to increasing the understanding of the impact of chatbot communication quality on brand experiences, a mostly unexplored research topic. The findings offer useful information that can help marketing practitioners by improving the communication quality of chatbot marketers would enhance the brand experiences of customers. The current study has strengthened the theory by demonstrating how the communication quality (credibility, openness, capability) of chatbot agents influences the brand experiences of customers. Brand managers and chatbot developers must pay particular attention and continuously strive to improve the communication quality of chatbots because message interaction increases the brand experience.

For practitioners and marketers across the globe, this research offers insights. The current study's empirical findings have produced several suggestions that could aid practitioners, system developers, and marketers in enhancing chatbot communication quality to provide a pleasant brand experience to customers. Researchers urge marketing experts to further increase their usage of AI-powered chatbots with good communication quality in marketing and customer service to enhance the brand experiences of customers. The researchers advise employing AI not only to enhance internal procedures but also as a tool (e.g., chatbots) to enhance customer service for low-complexity tasks, shifting people's resources to other business activities. Further, the brand name must be ingrained in consumers' minds by manufacturers so that when they ask for aid, they will remember to include it in their inquiry from an assistant virtual.

The theory has been strengthened by the current research, which discovered that the communication quality (credibility, openness, capability) of chatbots can increase the favourable brand experiences of customers. The usage of AI-based chatbots, for enhancing brand experience, will have helpful implications for businesses. However, several topics have not been thoroughly covered in the current study and may be considered in subsequent research.

4.5 LIMITATIONS

It's crucial to keep in mind that there are limitations to any study that offer interesting directions for further research. The economics, privacy, and democracy could all be harmed by digital helpers. Even if the technology exists for a neutral virtual assistant, if market forces and dynamics go unchecked, they may produce a powerful and dishonest digital assistant. To reduce the risks and protect the freedom and interests of consumers, regulators and legislators are taking action. This study was time and money constrained, despite the significant discoveries and insights it provided for the field of marketing. While there are many opportunities for AI to enhance people's lives and the growth of economies and communities, it also presents several technological concerns. Future researchers could investigate the connection between brand chatbot usage and actual consumer purchases.

REFERENCES

Alrubaian, M., Al-Qurishi, M., Alamri, A., Al-Rakhami, M., Hassan, M. M., & Fortino, G. (2019). Credibility in online social networks: A survey. *IEEE Access, 7*, 2828–2855. 10.1109/ACCESS.2018.2886314

Alsubayhay, A. M. S., Salam, M. S. H., & Mohamed, F. B. (2022). A review on approaches in Arabic chatbot for open and closed domain dialog. *International Journal of Advanced Computer Science and Applications, 13*(11). 10.14569/IJACSA.2022.0131117

Aw, E. C. X., Tan, G. W. H., Cham, T. H., Raman, R., & Ooi, K. B. (2022). Alexa, what's on my shopping list? Transforming customer experience with digital voice assistants. *Technological Forecasting and Social Change, 180*. 10.1016/j.techfore.2022.121711

Balaji, M., & Yuvaraj, N. (2019). Intelligent chatbot model to enhance the emotion detection in social media using bi-directional recurrent neural network. *Journal of Physics: Conference Series, 1362*(1), 012039. 10.1088/1742-6596/1362/1/012039

Bawack, R. E., Wamba, S. F., & Carillo, K. D. A. (2021). Exploring the role of personality, trust, and privacy in customer experience performance during voice shopping: Evidence from SEM and fuzzy set qualitative comparative analysis. *International Journal of Information Management, 58*. 10.1016/j.ijinfomgt.2021.102309

Baabdullah, A. M., Alalwan, A. A., Algharabat, R. S., Metri, B., & Rana, N. P. (2022). Virtual agents and flow experience: An empirical examination of AI-powered chatbots. *Technological Forecasting and Social Change, 181*. 10.1016/j.techfore.2022.121772

Camilleri, M. A. (2022). E-commerce websites, consumer order fulfillment and after-sales service satisfaction: the customer is always right, even after the shopping cart check-out. *Journal of Strategy and Management, 15*(3), 377–396. 10.1108/JSMA-02-2021-0045

Chakraborty, U., & Bhat, S. (2018). Credibility of online reviews and its impact on brand image. *Management Research Review, 41*(1), 148–164. 10.1108/MRR-06-2017-0173

Følstad, A., & Brandtzaeg, P. B. (2020). Users' experiences with chatbots: Findings from a questionnaire study. *Quality and User Experience, 5*(1). 10.1007/s41233-020-00033-2

Følstad, A., & Taylor, C. (2021). Investigating the user experience of customer service chatbot interaction: A framework for qualitative analysis of chatbot dialogues. *Quality and User Experience, 6*(1). 10.1007/s41233-021-00046-5

Fotheringham, D., & Wiles, M. A. (2022). The effect of implementing chatbot customer service on stock returns: An event study analysis. *Journal of the Academy of Marketing Science*. 10.1007/s11747-022-00841-2

Grewal, D., Guha, A., Satornino, C. B., & Schweiger, E. B. (2021). Artificial intelligence: The light and the darkness. *Journal of Business Research, 136*, 229–236. 10.1016/j.jbusres.2021.07.043

Hair, J. F., Risher, J. J., Sarstedt, M., & Ringle, C. M. (2019). When to use and how to report the results of PLS-SEM. *European Business Review, 31*(1), 2–24. 10.1108/EBR-11-2018-0203

Huang, M.-H., & Rust, R. T. (n.d.-a). *A strategic framework for artificial intelligence in marketing*. 10.1007/s11747-020-00749-9/Published

Haugeland, I. K. F., Følstad, A., Taylor, C., & Alexander, C. (2022). Understanding the user experience of customer service chatbots: An experimental study of chatbot interaction design. *International Journal of Human Computer Studies, 161*. 10.1016/j.ijhcs.2022.102788

Huang, W., Hew, K. F., & Fryer, L. K. (2022). Chatbots for language learning—Are they really useful? A systematic review of chatbot-supported language learning. *Journal of Computer Assisted Learning, 38*(1), 237–257. 10.1111/jcal.12610

Jansom, A., Srisangkhajorn, T., & Limarunothai, W. (2022). How chatbot e-services motivate communication credibility and lead to customer satisfaction: The perspective of Thai consumers in the apparel retailing context. *Innovative Marketing, 18*(3), 15–27. 10.21511/im.18(3).2022.02

Johannsen, F., Schaller, D., & Klus, M. F. (2021). Value propositions of chatbots to support innovation management processes. *Information Systems and E-Business Management, 19*(1), 205–246. 10.1007/s10257-020-00487-z

Khadija, A., Zahra, F. F., & Naceur, A. (2021). AI-powered health chatbots: Toward a general architecture. *Procedia Computer Science, 191*, 355–360. 10.1016/j.procs.2021.07.048

Klaus, P., & Zaichkowsky, J. L. (2022). The convenience of shopping via voice AI: Introducing AIDM. *Journal of Retailing and Consumer Services, 65*. 10.1016/j.jretconser.2021.102490

Kopalle, P. K., Gangwar, M., Kaplan, A., Ramachandran, D., Reinartz, W., & Rindfleisch, A. (2022). Examining artificial intelligence (AI) technologies in marketing via a global lens: Current trends and future research opportunities. *International Journal of Research in Marketing, 39*(2), 522–540. 10.1016/j.ijresmar.2021.11.002

Kushwaha, A. K., Kumar, P., & Kar, A. K. (2021). What impacts customer experience for B2B enterprises on using AI-enabled chatbots? Insights from big data analytics. *Industrial Marketing Management, 98*, 207–221. 10.1016/j.indmarman.2021.08.011

Liu-Thompkins, Y., Okazaki, S., & Li, H. (2022). Artificial empathy in marketing interactions: Bridging the human-AI gap in affective and social customer experience. *Journal of the Academy of Marketing Science.* 10.1007/s11747-022-00892-5

Meichan, Li, & Wang, R. (2023). Chatbots in e-commerce: The effect of chatbot language style on customers' continuance usage intention and attitude toward brand. *Journal of Retailing and Consumer Services, 71*, 103209, ISSN 0969-6989 , 10.1016/j.jretconser.2022.103209

Moore, S., Bulmer, S., & Elms, J. (2022). The social significance of AI in retail on customer experience and shopping practices. *Journal of Retailing and Consumer Services, 64*. 10.1016/j.jretconser.2021.102755

Mustak, M., Salminen, J., Plé, L., & Wirtz, J. (2021). Artificial intelligence in marketing: Topic modeling, scientometric analysis, and research agenda. *Journal of Business Research, 124*, 389–404. 10.1016/j.jbusres.2020.10.044

Oh, Y. J., Zhang, J., Fang, M. L., & Fukuoka, Y. (2021). A systematic review of artificial intelligence.

Pallant, J. L., Karpen, I. O., & Sands, S. J. (2022). What drives consumers to customize products? The mediating role of brand experience. *Journal of Retailing and Consumer Services, 64*. 10.1016/j.jretconser.2021.102773

Qi, X., & Kuik, S. (2022). Effect of word-of-mouth communication and consumers' purchase decisions for remanufactured products: An exploratory study. *Sustainability, 14*(10), 5963. 10.3390/su14105963

Riegger, A. S., Merfeld, K., Klein, J. F., & Henkel, S. (2022). Technology-enabled personalization: Impact of smart technology choice on consumer shopping behavior. *Technological Forecasting and Social Change, 181*. 10.1016/j.techfore.2022.121752

Ringle, C. M., & Sarstedt, M. (2016). Gain more insight from your PLS-SEM results: The importance-performance map analysis (October 31, 2015). *Industrial Management & Data Systems, 116*(9), 1865–1886

Romero-Brufau, S., Wyatt, K. D., Boyum, P., Mickelson, M., Moore, M., & Cognetta-Rieke, C. (2020). What's in a name? A comparison of attitudes towards artificial intelligence (AI) versus augmented human intelligence (AHI). *BMC Medical Informatics and Decision Making, 20*(1). 10.1186/s12911-020-01158-2

Schmitt, D. P., Realo, A., Voracek, M., & Allik, J. (2008). Why can't a man be more like a woman? Sex differences in Big Five personality traits across 55 cultures. *Journal of Personality and Social Psychology, 94*(1), 168.

Steinhoff, L., Arli, D., Weaven, S., & Kozlenkova, I. V. (2019). Online relationship marketing. *Journal of the Academy of Marketing Science, 47*(3), 369–393. 10.1007/s11747-018-0621-6

Sung, E. (Christine), Bae, S., Han, D. I. D., & Kwon, O. (2021). Consumer engagement via interactive artificial intelligence and mixed reality. *International Journal of Information Management*, *60*. 10.1016/j.ijinfomgt.2021.102382

Vlačić, B., Corbo, L., Costa e Silva, S., & Dabić, M. (2021). The evolving role of artificial intelligence in marketing: A review and research agenda. *Journal of Business Research*,128, 187–203. 10.1016/j.jbusres.2021.01.055

Volkmar, G., Fischer, P. M., & Reinecke, S. (2022). Artificial intelligence and machine learning: Exploring drivers, barriers, and future developments in marketing management. *Journal of Business Research*, *149*, 599–614. 10.1016/j.jbusres.2022.04.007

Witell, L., Kowalkowski, C., Perks, H., Raddats, C., Schwabe, M., Benedettini, O., & Burton, J. (2020). Characterizing customer experience management in business markets. *Journal of Business Research*, *116*, 420–430. 10.1016/j.jbusres.2019.08.050

Xu, Y., Zhang, J., & Deng, G. (2022). Enhancing customer satisfaction with chatbots: The influence of communication styles and consumer attachment anxiety. *Frontiers in Psychology*, *13*, 902782. 10.3389/fpsyg.2022.902782

Yadav, A., & Dhanda, N. (2023). An Empirical Study of Design Techniques of Chatbot, a Review. In P. K. Singh, S. T. Wierzchoń, S. Tanwar, J. J. P. C. Rodrigues, & M. Ganzha (Eds.), *Proceedings of Third International Conference on Computing, Communications, and Cyber-Security* (Vol. 421, pp. 139–151). Singapore: Springer Nature. 10.1007/978-981-19-1142-2_11

Yen, C., & Chiang, M. C. (2021). Trust me, if you can: A study on the factors that influence consumers' purchase intention triggered by chatbots based on brain image evidence and self-reported assessments. *Behaviour and Information Technology*, *40*(11), 1177–1194. 10.1080/0144929X.2020.1743362

Zhang, C., & Lu, Y. (2021). Study on artificial intelligence: The state of the art and future prospects. *Journal of Industrial Information Integration*, *23*. 10.1016/j.jii.2021. 100224

Zarantonello, L., & Schmitt, B. H. (2013). The impact of event marketing on brand equity: The mediating roles of brand experience and brand attitude. *International Journal of Advertising*, *32*(2), 255–280.

5 Making the Impossible Possible

Tata Nano

Supriya Lamba Sahdev
Department-Management, Assistant Professor and
International Programs Manager ISBR Business School,
Bengaluru, Karnataka, India

Ahdi Hassan
Researcher Global Institute of Research and Scholarship
Amsterdam, Netherland

Chitra
Associate Professor, Symbiosis Centre for Management
Studies (SCMS Noida), Uttar Pradesh

Hemanshu Yadav
PGDM in Product Management, Research Scholar,
ISBR Business School, Bengaluru, Karnataka, India

5.1 INTRODUCTION TO FRUGAL INNOVATION

Many manufacturing sectors have taken an interest in the idea of frugal innovation because they want to offer straightforward and affordable products to developing markets where consumer needs differ greatly from those of developed nations. In order to provide new, practical insights for practitioners, the following requests have been made for the field of global business research beyond Western industrialized economies to include emerging and underdeveloped economies.

Innovation is characterized as the application of concepts to the creation of goods or services (Schumpeter & Backhaus, 2003). It is critical to generating economic growth, and even a modest rise over time can produce income differences that are very visible (Ahlstrom, 2010). Companies are putting more emphasis on creating solutions for clients with limited resources (Viswanathan & Sridharan, 2009). The key to satisfying underserved clients is to use integrated design solutions and integrated economical innovations to address their unmet demands.

The increase in inventions coming from growing countries in the recent past dispels misconception that only ideas originate in the West and then spread to the

DOI: 10.1201/9781003365525-5

rest of the world (Petrick & Juntiwasarakij, 2011). Perhaps the most recent idea to be discussed in the management research of such a model is frugal innovation (FI). FI is regularly cited by researchers as a potential option for helping the underprivileged who cannot afford conventional products (Hossain & Sarkar, 2021). The benefit of frugal innovation is that it makes moderate customers active purchasers rather than passive recipients of contributions (Vadakkepat et al., 2015).

Frugal innovation, also referred to as Jugaad innovation and Gandhian innovation, is defined as "a resource-scarce solution (i.e., product, service, process, or business model) that is designed and implemented despite financial, technological, material, or other resource constraints, whereby the outcome is significantly less expensive than competitive offerings (if available) and is good enough to meet the basic needs of customers who would otherwise remain underserved" (Hossain et al., 2016). The term "bottom of the pyramid" (BoP) refers to those with daily incomes significantly lower than $2.50; nevertheless, the terms "Jugaad" and "Gandhian" innovation are geographically exclusive to India. BoP has the strongest connection even though it is a more particular idea than frugal innovation (Hossain, 2018). The Hindi word "jugaad" refers to a creative upgrade that calls on talent and imagination (Radjou et al., 2012). In addition, jugaad describes improvization and innovative approaches to common problems that make use of fresh applications of the resources at hand. Constrained-based innovation, which is even more encompassing than "frugal innovation," also encompasses ideas like "reverse innovation," "blowback innovation," and "trickle-up innovation" (Agarwal et al., 2017).

Frugal innovation, which is sometimes referred to as "frugal engineering," is a process where the complexity of a product is decreased along with the price of a good and its production. Recently there has been an increase in the importance of frugal innovation throughout the world, particularly in emerging or developing nations where people have less purchasing power and seek affordable, high-quality utilities. People in underdeveloped nations prefer to choose more affordable goods over more expensive ones. Frugal innovation is therefore extremely important in these nations. The phrase "frugal innovation" simply describes a method whereby an innovation's complexity is minimized in order to lower production costs and offer items at lower prices in a high-volume market.

Frugal engineering goes beyond simply using "cheap" engineering to increase profits through supplier negotiations. Instead, it is a way of thinking that breathes new life into the idea of product development. Cost control is crucial to the process, but frugal innovation attempts to eliminate wasteful spending during the design stage rather than just reducing current expenditures. This strategy recognizes that simply taking away features from current goods to reduce their price for growing markets is ineffective since consumers in these regions have particular requirements and desires that cannot be satisfied by only altering some attributes. Additionally, the upfront costs of items manufactured in advanced nations are frequently over expensive for permitting competitive pricing in developing markets (Figure 5.1).

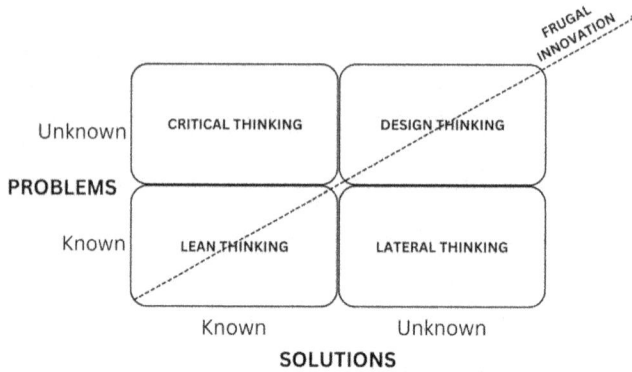

FIGURE 5.1 Describes the approach of frugal innovation with problem and solution context.

Source: Design Your Thinking by Pawan Soni.

5.2 INDIA AND FRUGAL INNOVATION

India, a country with a population of over 139.38 crores (1.39 billion), has a longstanding culture of maximizing the use of limited resources through re-engineering. The phrase "jugaad," which essentially means finding a simpler solution to a problem, is frequently used to describe frugal innovation in India. India stands 46th among the 132 economies featured in Global Innovation Index. While addressing the Indian diaspora in Germany, Prime Minister Narendra Modi said India has made significant progress and is the global leader in the "Way of Innovation." Atal Innovation Mission was launched in 2016; the plan was developed to promote innovation by developing new programmes and policies to assist growth in a number of economic sectors. It offers a platform for the collaboration of numerous entrepreneurs. India, which has a total of 25 innovation centres, has been ranked second in the world and first in Asia for the establishment of new innovation centres. In addition, the Commerce and Industry Ministry reports that as a result of the Indian government's efforts to improve the nation's intellectual property rights (IPR) system, the total number of patent applications in India increased from 42,763 in 2014–2015 to 66,440 in 2021–2022.

Even well-known businesses like Godrej, Tata, and Nokia have employed frugal engineering to build products that have found success in the Asian markets. Asian nations have inserted frugal innovation as a vital component of their extremely stimulated industrial industries. Spandan, Sewage Pipe Homes, Multi-Harvester, Tupik Bed AC, Mitti Cool, Jaipur Foot, and Voice Box are a few unusual technologies from the past ten years that have improved and safeguarded people's lives. Companies like HUL and P&G have been selling shampoo and laundry detergent in tiny packets in urban India for more than 20 years in an effort to appeal to customers who were on a low budget. This may have been the start of the thrifty innovation trend. The introduction of sachet packs, not only increased the availability of these products on the market, transforming the FMCG sector, but they also made it possible for businesses with constrained space to store merchandise more efficiently without having to add extra space for shampoo bottle displays.

5.3 TATA NANO

Globalization has made the world a smaller place since liberalization in 1991. Markets prioritize industrialized goods while ignoring innovation. The Tata Nano, manufactured by India's largest automaker, stands out as an exception in this situation. The car is the most reasonably priced in the entire world and best satisfies the needs of the middle- and lower-income families that make up the majority of the Indian market. It was developed using the frugal innovation technique. This car is a perfect example of modern India's ingenuity. This was made possible by the company's understanding of the needs of its customers in the current business climate and its prudent fulfilment of those needs through significant R&D in the creation of the Tata Nano automobile.

The demand for inexpensive vehicles was expanding when the Tata Nano debuted in 2009. Even though there was a sizable middle class and about 50 million people who rode motorcycles as their major mode of family transportation, individual automobile ownership was still uncommon. The Tata Group's chairman, Ratan Tata, asserts that he was motivated to create an affordable vehicle for every Indian after witnessing a family with children travelling through congested traffic on a motorcycle. The $2,000 Tata Nano automobile and the cheapest water filter in the world, the Tata Swach, that doesn't need energy or running water, are just a couple of the company's famously affordable products. Ratan Tata, the group chairman, created the Nano with the intention of giving families and small businesses, who traditionally relied on auto-rikshaws and railroads, a means of transportation. The Nano was at the time India's most affordable automobile, costing INR 1 Lakh, or around $1,400. The Tata Nano research team's five-year R&D project, which began in 2003, produced the Nano. While upholding the quality of every component used in the construction of the car, such as the gear shifter, windscreen washer system, steering wheel, wheels and tyres, etc., Tata Motors and its suppliers consistently fought to keep prices low. Engineers were required to stick to the low-price guarantee by being as frugal as they could while designing.

5.3.1 TACTICAL METHOD

Tata Motors made engineering advancements while creating the Nano that had an impact on both the finished vehicle and the entire organizational ethos. Creating a high-quality car for Rs. 1 lakh (about $1,400) that would appeal to bourgeois households and eventually supplant market for two-wheelers was the goal of the entire process, from team formation to product delivery. Before production could begin, It took over four years to complete the development plan. Under the direction of design leader Girish Wagh, the engineering team swelled to 500 individuals, with each engineering team responsible for a different Nano component. Ultimately, a 20-person production team was put together after a rigorous assessment and interviewing procedure. To build a diversified workforce that would produce original and innovative ideas, Tata Motors hired apprentice engineers from IIT Kharagpur, Jadavpur University, and other institutions.

Tata Motors was willing to take risks, try new approaches, and involve its business partners in the product design process for the Nano. The management group opted to carry out "concurrent engineering in real time" rather than just giving them technical specifications; early product development should involve component suppliers. Suppliers were urged to review the current work and make their own insightful recommendations that could help further encourage a frugal outlook.

For instance, Tata Motors and the German business Bosch worked together to develop a new engine control system (EMS). By removing pointless activities, Bosch was able to shrink the programme's size and expand the amount of micro-controller computing power that was accessible. To resist the rigours of Indian roads, the Madras Rubber Factory in India created harder rear tyres. Additionally, Tata collaborated with Japanese auto component producer Denso to create a novel windscreen a single-wiper system as opposed to two-wiper systems, saving the Nano team extra expenses. In order to get access to additional revenue streams, more than half of the supplied components were given permission to be enhanced as patented technology of the specific provider.

5.3.2   PRACTICAL METHOD

In an effort to maintain a strong feeling of thriftiness and question conventional vehicle manufacturing processes, Tata's efforts to produce an economical car included leaving out equipment such as radios, power steering, power brakes, and air conditioning. The Nano may be distributed, assembled, and serviced by local business owners as kits since it is made up of parts that can be sent separately to various locations and installed there. Tata Motors was able to submit 34 patents for the creation of the Nano, which was intended to redefine economic quality in the automotive industry, thanks to Ratan Tata's foresight and a group of committed engineers.

5.3.3   MOUNTING AND SMALL WHEELS

The Nano's small, lightweight wheels were specifically intended to enhance per-formance, economy, and riding comfort. It employed press-steel "style-in-steel" wheels, which look like alloys but are almost as expensive as standard steel wheels. As designers tried to review and enhance options that had been customary in the building of cars, in order to reduce expenses, these wheels were likewise attached with just three lugs. Due to the car's modest weight and small wheels, power steering is probably not going to be necessary, which further minimizes complexity and expenses. The design team kept the driving wheels thicker while giving the front wheels a somewhat thinner specification. The automobile should have a tiny understeer at the limit because of the wider track up front, which changes the effect and serves as a safety element. To put it simply, the engine of the Nano causes a weight bias toward the rear. This alone can be a problem when cornering quickly because the rear end starts to act somewhat pendulum-like, tugging the car off the line. Depending on how well the design is prepared, the engineered understeer should balance and neutralize the automobile. The smaller tyres also resulted in cost savings on extra rubber. Not to mention the closed hatch, which means there won't be any beading, hinge, or lock costs, and the entire panel can be a reasonably affordable addition to the monocoque, improving the chassis' strength without costing more. Non-opening hatches also saved money on beading, hinges, and lock.

5.3.4 ENGINE

The 624cc four-stroke parallel-twin engine is covered by ten of Tata's patents from the Nano project. Together with Tata, the German manufacturer Bosch created a new engine management system. The motor with a single counter balancer has been dubbed a world first for an automobile application. The four-speed transaxle is coupled directly to the engine, which is positioned transversely behind the back seats and in front of the rear axle line. Nano is similar to the original Fiat 500, which has a rear-mounted engine to aid maximize cabin space, making it another illustration of a technologically sophisticated economic car. The driveshaft is removed with the rear-engine, rear-wheel drive configuration, which reduces costs.

After testing over 32 different versions of the best driveshaft for a year, GKN Driveline India, a division of the world's largest auto parts manufacturer GKN, produced the driveshaft. The company enlisted designers from its French and Italian subsidiaries to create a simple and affordable design. For Nano's rear-wheel drive system, GKN developed a smaller diameter shaft, which made it lighter and utilized less material. Additionally, the powertrain is located beneath the back seats rather than in the front. Because the hatch does not open, the back seats must be moved forward in order to reach the engine. Tata may have intentionally designed an engine that is relatively practical and efficient considering the difficulty of access to it due to the layout.

5.3.5 INTERIOR

The Nano's design is focused on comfort and usability. After going through multiple versions to cut costs, the seating was eventually created as a traditional seat with a metal structure and a headrest built into the seat. It also features a centralized instrumentation that is easily visible to all passengers, creating a more open feeling in the car. The large plastic component does not need to be adapted for left- and right-hand drives thanks to the central metre set. It includes a basic set of lights, an analogue speedometer, an odometer, and a digital fuel gauge.

Customers had the option to add features like airbags, anti-lock brakes, a radio, and air conditioning to their models, but these features were not included as standard. Not including these non-essential features reduced the average cost for the basic models. The smooth curves, two large storage compartments, and ample storage space make the cabin feel light and spacious, and they do away with the requirement of a glove box. The curving edges might also increase collision protection.

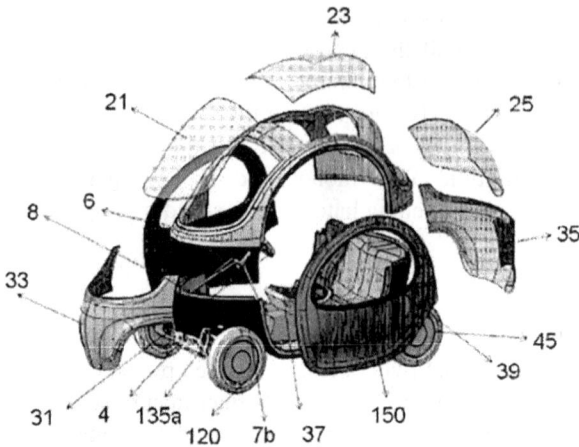

FIGURE 5.2   Patents of Tata Nano.

5.3.6   EXTERIOR, SIZE, AND TRANSMISSION

The whole body of the Nano is steel. Instead of being painted, it has a straightforward steel platform and steel body panels. Under the nose is where you'll find the filler cap. This saves the cost of drilling a hole in the body for a filler cap, and drivers are already familiar with these because they are the typical filling procedures for CNG and LPG vehicles. The Nano does not include a left-side mirror. The Maruti 800, the next smallest car marketed in India, is a little over ten feet longer than the Nano, which is about 20% smaller overall. However, the Nano has a larger cabin than the Maruti. Reduced cost and weight resulted from fewer parts and weights. It came as a small shock that the Nano features a four-speed manual transmission. The CVT is small, straightforward, and makes efficient use of the engine's power.

5.3.7   PATENT

Tata Motors has applied for 31 design patents and 37 technology patents with an emphasis on the technological development of the Tata Nano. The technologies covered by the patents include material technology, manufacturing technology, and vehicle technology. Tata Nano has cutting-edge technologies, as shown by these patents. The technologies are designed to produce goods that are incredibly affordable and highly reliable (Figure 5.2).

5.4   TATA NANO AND THE INTERNATIONAL MARKET

Data from the *World Street Journal* indicate that the U.S. market has a lot of potential for small and environmentally friendly automotive manufacturers. The global financial crisis and growing awareness of the challenges posed by large cars have impacted consumer behaviour in the U.S. market. Customers bought cars at reduced prices as a result of the recession, and this factor encouraged tiny car manufacturers to enter this market. Tata Motors would therefore find it simple to take a sizable market share of the U.S. auto markets with their Tata Nano model.

According to Noronha (2005), Tata Motors would have a competitive advantage if it established a distribution channel for the Tata Nano as the Tata Group has a long history of integration with the U.S. market dating back to the early 1900s and offers a wide range of goods and services. The company's stock is traded on the New York Stock Exchange and has 80 offices across the country with 10,000 employees. However, Nano's main competitive edge is its ability to minimize costs, which will enable the business to grow in the post-recessionary period. It's likely that the firm may upgrade the electric version of the vehicle, making it the least expensive electric vehicle with a 130 km range and lithium-ion batteries.

5.5   FRUGAL INNOVATION AND THE WORLD

Innovations are typically created in developed nations and subsequently exported to underdeveloped nations. However, there has been a noticeable rise in creativity recently in situations with limited resources, notably in developing nations. Consumers in developing nations are a new target market for global firms because of their reduced purchasing power and distinctive shopping habits. In order to create goods that address local needs, Western firms frequently work with their affiliates in developing nations. This collaboration enables them to combine cutting-edge knowledge with local expertise to provide effective solutions. Although there is little research on how to establish innovation capabilities for these clients in developing nations, there is a potential for a new market in developing countries by creating innovations for customers who have previously been underserved. Innovation plays a major role in driving economic growth, which can lead to significant income differences over time, even if the economic growth seems small at first. Innovation is seen as a crucial engine of economic growth, and over time, even slight differences in economic growth can lead to significant differences in income.

5.5.1   EXAMPLES OF FRUGAL INNOVATION AT THE GLOBAL-LEVEL LOGIQ BOOK

Another famous instance of a low-cost innovative, the portable ultrasound machine, is the product made by GE called the Logiq Book, which was created with rural China in mind. While having fewer functions than traditional ultrasound devices, this computer-based laptop gadget offers further benefits that are essential to its rural target markets. The Logiq book is substantially lighter, smaller, and only offers the most basic capability, despite the fact that the cost of conventional ultrasound systems has been cut by almost 80%. Due to its portability, it saves rural patients the trouble of travelling to distant cities for diagnosis and treatment.

5.5.2   NOKIA 1100

One well-known example of thrifty innovation is the Nokia 1100. It was released in 2003 as a mobile phone for developing countries and is regarded as one of the most popular models worldwide. Nokia reduced the device's components to the bare minimum required by low-income consumers in order to keep the price as cheap as possible while also adding helpful features like a flashlight and a potent battery that made it durable and dustproof. Despite being basic and only capable of doing necessary chores, it was highly regarded by customers.

5.5.3   M-KOPA

It is a pre-configured solar energy system for homes. What most intrigues me, though, is the device itself and the unconventional business strategy employed to keep costs low for all clients. The target market for the product, Kenyans, would not be able to afford the $200 price tag. What makes it so special is that anyone may use a mobile phone to make daily micropayments of 45 cents for as little as a $25 commitment. However, they were developed through frugal innovation using a mobile payment system.

5.5.4   Oggun Tractor

Cleber, a small company based in Alabama, used the use of frugal innovation ideas to meet Cuba's agricultural demands. In Cuba, small and cooperative farmers cultivate the majority of agricultural land and face challenges such as a lack of resources, outdated equipment, and inefficient distribution and transportation. Additionally, the low average monthly salary of Cuban customers, combined with an increase in tourism and demand for high-quality products, made these problems worse. Cleber recognized that the influx of American tourists could have negative consequences for Cuba, and wanted to offer competitively priced, easy-to-use products as an alternative to current Cuban products. To achieve this, the company developed the Oggun tractor, a compact and lightweight zero-turn tractor powered by a 22-horsepower gasoline engine. The tractor was designed to be easy to use and efficient, with an open operator's seat that provides unobstructed views of the ground.

5.5.5   Jaipur Foot, Knee, and Limb

Approximately 100,000 amputees worldwide now live better lives thanks to Jaipur Foot, Knee, and Limb, which has provided many people freedom of movement and dignity. Their plan was simple, and the price for a flexible prosthetic foot was a mere $45 (USD). However, the "Jaipur Foot," as it was known by some of the most marginalized members of society, had amputees for the first time who lowered the cost of prosthetics. In contrast to the majority of prosthetics created at the time, which were made of carbon fibre, the "Jaipur Foot" was made of rubber, plastic, and wood. Employees could finally purchase prostheses since the production cost was reduced to $45 as a result. As a result, manufacturing costs sharply decreased. They were not only far more versatile and less expensive than their western counterparts, but also did not call for the use of shoes. They are much more comfortable and practical for daily activities because many people wear them when walking barefoot.

Frugal innovation, which involves developing products or services that are affordable and efficient, has become increasingly important in resource-limited situations, such as during the COVID-19 pandemic. While several approaches to frugal innovation focus on sustainability, there is no widely accepted theory or dominant paradigm in this area of research, which is still relatively new and requires further study. There is also a significant potential for frugal innovation to target underserved markets, including customers with modest incomes in both developed and emerging countries, as income inequality and poverty rates continue to increase.

5.6 SCOPE OF FRUGAL INNOVATION AT THE GLOBAL LEVEL

Frugal innovation has a wide range of global applications, including in the following areas:

- Healthcare: Frugal innovation can be used to develop affordable healthcare technologies, such as portable diagnostic devices and telemedicine platforms, which can improve access to healthcare for people in underserved markets.
- Clean energy: Frugal innovation can be applied to the development of cost-effective clean energy sources like wind turbines and solar panels, which can help to reduce greenhouse gas emissions and address climate change.
- Education: Frugal innovation can be used to develop affordable educational technologies, such as online learning platforms and low-cost tablets, which can improve access to education for people in underserved markets.
- Transportation: Frugal innovation can be used to develop affordable and efficient transportation solutions, such as electric bikes and shared mobility platforms, which can help to reduce congestion and improve access to transportation in urban areas.
- Agriculture: Frugal innovation can be used to develop affordable and sustainable farming technologies, such as precision irrigation systems and low-cost greenhouse systems, which can improve food security and reduce water use.
- Social issues: Frugal innovation can be used to create cost-effective solutions for social problems, such as homelessness or food insecurity, in developed countries.

5.7 GLOBAL IMPORTANCE

First, it can help address global challenges related to access and affordability. By developing products and services that are accessible and affordable to people in emerging markets or with limited resources, frugal innovation can help to narrow the divide between the wealthy and the impoverished and improve the quality of life for people around the world. Second, frugal innovation can drive economic growth and development in emerging markets. By developing new offerings that address the needs of groups that have previously been neglected, frugal innovation can help stimulate local economies and create new opportunities for economic development.

Third, frugal innovation can help address global sustainability challenges. By designing products and services that use fewer resources and have a smaller environmental footprint, frugal innovation can help reduce waste and resource consumption and contribute to a more sustainable future. Overall, frugal innovation has the capacity to significantly impact positive effect on global challenges related to access, affordability, economic development, and sustainability.

5.8 CHALLENGES AND ACTION

Tata encountered numerous obstacles on the route to developing Nano, which was not an easy journey. Tata's initial proposals to locate the factory in West Bengal were rejected by opposition parties, leading to a last-minute change in location to Sanand. When the company first started operating, there weren't enough automobiles being produced to satisfy the influx of requests. The orders eventually stopped. Additionally, the safety of Nano was questioned after a few cars caught fire, which negatively impacted consumer opinion. Moreover, the sales continued to decline amid a downturn in the Indian auto industry brought on by rising loan rates and gasoline costs. Tata Motors needed to design a car that could serve as a true family vehicle with four doors, rather than just a motorized quad bike. In order to compete with the Maruti 800, which was the most inexpensive car available at the time, they also set a target selling price of around $1,400.

Nano received a makeover from TML with the launching of variants with stronger and more fuel-efficient engines. The new Nano versions are priced between Rs 1.75 lakh and Rs 3.25 lakh due to a restructuring of the pricing. Instead of just selling a cheap car, they are now thinking of selling the entire vehicle. Tata plans to offer a range of additional features, such as four engine options, two glove compartments, keyless entry, a dashboard audio system with Bluetooth connectivity, a larger steering wheel, and remote central locking, for an additional cost. The launch of the power steering and AC variant accounts for 60% of Nano sales. To cater to younger clients, Tata has also redesigned Nano's appearance by offering it in a variety of vivid colours. The Nano is still relatively inexpensive in the business hatchback market even with the addition of extra features to boost value, with a starting price of 1.75 lahks (about $2,400).

And as for frugal innovation challenges, one challenge is the lack of access to resources and funding. Many emerging markets and low-income communities may have limited access to financial resources, technology, and expertise, which can make it difficult to develop and implement frugal innovations. Another challenge is the need to balance affordability and quality. Frugal innovations are designed to be affordable and accessible to people with limited resources, but they also need to be of high-enough quality to meet the needs and expectations of users. Striking the right balance between affordability and quality can be challenging. In addition, frugal innovations may face challenges related to cultural and social norms. For example, some frugal innovations may not be accepted or adopted in certain communities due to cultural or social barriers. Finally, frugal innovations may face challenges related to regulatory and policy environments. For example, some countries may have regulatory barriers or lack the necessary infrastructure to support implementing and scaling frugal innovations.

Overall, these challenges can make it difficult to develop and implement frugal innovations at a global level, but with the right strategies and approaches, it is possible to overcome these challenges and drive meaningful impact and innovation.

5.9   CONCLUSION

The Tata Nano is an idea that can only be realized if it is supported by a sincere foresight. The creation of Nano reinforced the notion that frugal innovation involves not fundamentally altering a product's engineering but also fundamentally altering organizational culture to provide more for less money. After the car's market debut, critics who had initially expressed doubts about the project's success gave Nano a favourable review.

The Nano rapidly became known as one of the most outstanding cases of innovative cost-effectiveness. The Ford Model T and Renault's "Logan" were the first to promote the idea of a cheap, environmentally friendly car to the automotive industry, but the Nano ultimately outperformed both of them in terms of low-cost engineering and market appeal, becoming the world's cheapest commercial vehicle. Tata made a significant achievement in manufacturing with the creation of the Nano, which was made from modular components that could be shipped separately and assembled at various locations. This modular concept has revolutionized the automotive industry and inspired many other car manufacturers to adopt innovative engineering techniques.

The Nano's success in engineering and economics has provided a foundation for the development of new products and the expansion into new markets. After clearing the European crash testing with flying colours, it intends to reach the African market. Given Nano's early success with cost-effective engineering, it has been rumoured that Tata is also developing hybrid and all-electric Nano models. The Nano has modernized its image while keeping its fundamental form in order to appeal to urban youngsters. With features including radio, Bluetooth, and air conditioning, Tata has introduced upgraded models such as the Nano Twist and Nano Max.

When it comes to solving global issues like access, affordability, economic growth, and sustainability, frugal innovation is a potent tool. It entails using the least amount of materials and financial resources possible to provide goods and services that are accessible and cheap to people in developing countries or with low resources. Frugal innovation has the potential to have a big positive impact on global concerns, but it also faces difficulties with funding and resource access, balancing affordability and quality, cultural and social norms, and regulatory and legislative contexts. Frugal innovation must be carefully planned and put into practice while taking into account the unique needs and difficulties of the target market to be successful on a global scale.

REFERENCES

Agarwal, N., & Brem, A. (2017). Frugal innovation-past, present, and future. *IEEE Engineering Management Review*, 45(3), 37–41.

Ahlstrom, D. (2010). Innovation and growth: How business contributes to society. *Academy of Management Perspectives*, 24(3), 11–24.

Bhatt, P., & Pathak, P. (2016). Indian frugal innovation and inclusive growth: An empirical study of impact on affordable healthcare. *Journal of Indian Business Research*, 8(4), 295–317.

Bhatt, P., & Pathak, P. (2019). From frugal to global: An exploratory study on Indian firms' pathways to global frugal innovation. *International Journal of Technology Management*, 81(1–3), 18–41.

Hossain, M. (2018). Frugal innovation: A review and research agenda. *Journal of Cleaner Production*, 182, 926–936.

Hossain, M., & Sarkar, S. (2016). Determinants of frugal innovation: A study of Indian manufacturing SMEs. *Journal of Manufacturing Technology Management*, 27(2), 217–241.

Hossain, M., & Sarkar, S. (2021). Frugal entrepreneurship: Profiting with inclusive growth. *IEEE Transactions on Engineering Management*.

IvyPanda (2023). International Marketing Plan for Tata Nano Quantitative Research (https://ivypanda.com/essays/international-marketing-plan-for-tata-nano/)

Petrick, I. J., & Juntiwasarakij, S. (2011). The rise of the rest: Hotbeds of innovation in emerging markets. *Research-Technology Management*, 54(4), 24–29.

Radjou, N., Prabhu, J., & Ahuja, S. (2012). *Jugaad innovation: Think frugal, be flexible, generate breakthrough growth*. John Wiley & Sons.

Schumpeter, J., & Backhaus, U. (2003). *The theory of economic development*. Schumpeter, J. A. (ed.) (pp. 61–116). Springer, Boston, MA.

Vadakkepat, P., Garg, H. K., Loh, A. P., & Tham, M. P. (2015). Inclusive innovation: Getting more from less for more. *Journal of Frugal Innovation*, 1(1), 1–2.

Viswanathan, M., & Sridharan, S. (2009). From subsistence marketplaces to sustainable marketplaces: A bottom-up perspective on the role of business in poverty alleviation. *Ivey Business Journal*, 73(2), 1–15.

6 AI in E-Commerce

Industry 4.0

Supriya Lamba Sahdev
Associate Professor and HOD Marketing, ISBR Business
School, Bengaluru, Karnataka, India

Navleen Kaur
Assistant Professor, Amity International Business School,
Amity University, Noida, India

Veera Shireesha Sangu
Research Associate, ISBR Business School, Bengaluru,
Karnataka, India

6.1 INTRODUCTION

6.1.1 INTRODUCTION TO A NEW TYPE OF INDUSTRIALIZATION: INDUSTRY 4.0

Industry 4.0 is reinventing the way companies produce, enhance, and distribute their products. Manufacturers are integrating the latest technologies, i.e., the Internet of Things (IoT), cloud computing, analytics, and artificial intelligence (AI).

- **IoT** – IoT refers to the collective network of connected devices that facilitates communication between devices and the cloud, e.g., smart grids or activity trackers.
- **Cloud Computing** – Cloud computing works by enabling devices over the Internet from remote physical servers, databases, and computers.
- **Analytics** – Analytics is the process of discovering, interpreting, and communicating significant patterns in data.
- **Big Data** – Big Data refers to the data with great variety, arriving in mass volumes with fast velocity.
- **AI** – Artificial intelligence refers to systems or machines that mimic human intelligence to perform tasks and can improve themselves based on the information they collect.

6.1.2 AI IN E-COMMERCE

AI in commerce is used to offer personalized recommendations based on past customer behaviour and lookalike customers. It is also being used to improve the

DOI: 10.1201/9781003365525-6

relationships between the sellers and customers using virtual assistant and chatbots, e.g., Amazon.

The main factor of Industry 4.0 is the involvement of AI in the digital industrialization, industries are focusing on improving their product consistency and productivity while trying to reduce their operating cost. In this chapter, we will focus on studying the relationship between humans, machines, and AI, and how this relationship can be used to enhance productivity with the available resources.

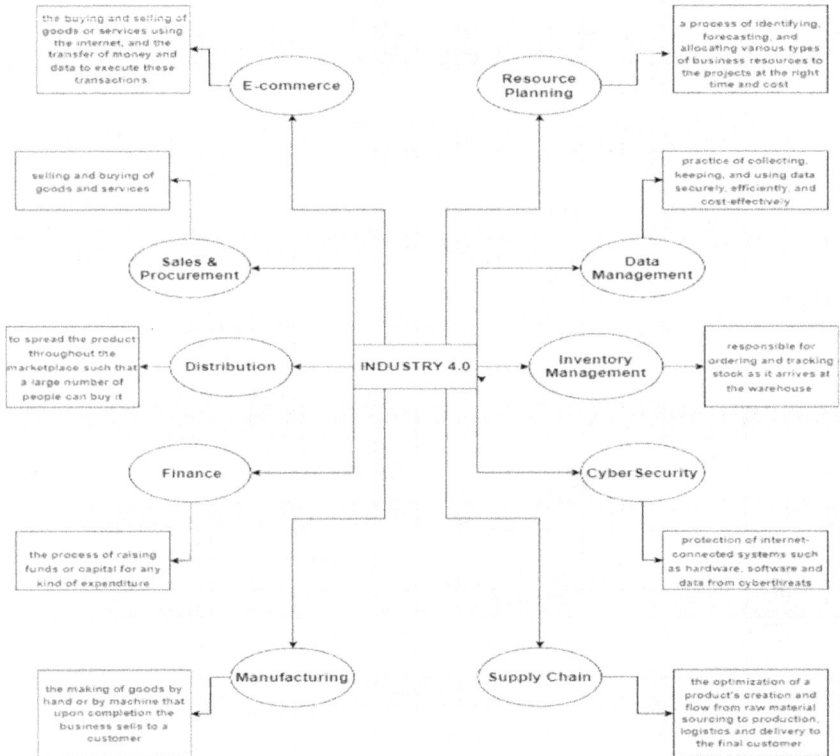

6.2 REVIEW OF LITERATURE

6.2.1 ARTIFICIAL INTELLIGENCE IN ER – METROPOLIAV

The main goal of this thesis is to study AI and how it is used in Enterprise Resource Planning (ERP). They used the existing literature and analyzed different publications and web articles about AI in ERP, i.e., sales forecasting from the historical data with the help of an SAP cloud application (SAC). (SAC is a data analysis tool that combines business intelligence, visualization, planning, and predictive functionalities in one solution.) They conclude that with the help of AI, companies can reach a new level of analytical productivity, i.e., customer service, sales forecasting, and predictive analysis.

6.2.2   ARTIFICIAL INTELLIGENCE IN DATABASE MANAGEMENT – IDEAS

According to this thesis, the integration of AI and database management systems (DBMS) play a crucial role in data processing. They counter the ever-growing data volumes and the need for fast analytical data processing. This integration needs access to large amounts of data for data processing, efficient management, and intelligent processing of data. The same model is being opted by the IT industries to create autonomous hardware and the first autonomous databases. In this thesis, they expect that by using methods of machine learning and sophisticated programme algorithms most of the maintenance will be done automatically. They also observe significant improvements in the hardware and software of complex engineered database machines.

6.2.3   AI-DRIVEN CYBERSECURITY – SPRINGER LINK

According to this thesis, AI is one of the key technologies of Industry 4.0 that can be used for the protection of Internet-connected systems. It uses deep learning, machine learning, the concept of natural language processing (ability to understand text and spoken words), knowledge representation (representing information about the world in a form that a computer system can use to solve complex tasks), or rule-based expert systems (prescribed knowledge-based rules to solve a problem) to prevent cyber threats, attacks, or unauthorized access. This security intelligence model can make the cybersecurity computing process automated and more intelligent than the conventional security systems.

6.2.4   ARTIFICIAL INTELLIGENCE IN INVENTORY MANAGEMENT – SCIENCE DIRECT

According to this thesis, reducing the wastage of goods and resources, and therefore cost, within a supply chain is a very challenging task because of the multiple variables involved. They reveal that unnecessary wastages are due to unnecessarily high inventory costs and shortage costs that are caused by errors in demand forecasting. These wastages can be avoided by using an artificial neural network that accurately forecasts demand. This model works on a multilayer feed-forward neural network (an interconnection of perceptron in which data and calculations flow in a single direction, from the input data to the outputs) with back propagation (algorithm that is designed to test for errors working back from output nodes to input nodes).

6.2.5   AI IN SUPPLY CHAIN MANAGEMENT – TANDFONLINE

In this thesis, they show how AI has shown great promise in improving decision making and analyzing business patterns, seeking information, and learning business phenomena to increase productivity. They think that AI has seen limited application in supply chain management (SCM). They explore the various fields in AI to solve the practical problems in SCM, such as managing customer experiences, managing supplies, maintaining quality, etc.

6.2.6 AI in Advanced Manufacturing – ASME Digital Collection

According to this thesis, the manufacturing processes are becoming increasingly complex, and connected. Recent developments have shown that AI and machine learning can help counter these problems by using advanced research tools that process vast amounts of manufacturing data (Big Data). This kind of model considers a wide variety of data from quality, process monitoring, diagnosis and prognosis, and supervisory control in human–robot relations.

6.2.7 AI in Distribution of Goods and Services – Science Direct

According to this thesis, the efficiency of the process of distribution of goods and services can be improved by using a data-driven machine learning system. This data-driven machine consists of a geocoding algorithm (i.e., precise calculation of the delivery address based on the transmitted information), a route calculation algorithm (solving the vehicle routing problem with time windows (VRPTW) with a random probability distribution), and distribution region optimization algorithms. These machines also monitor vehicle traffic using smart tracking devices.

6.2.8 AI in Sales and Procurement – Informs Pub Online

In this thesis, they talk about the ways in which AI affects sellers in selling and buyers in procurement. They talk about the impact of automation, i.e., buyers using a chatbot to inquire about prices instead of asking in person. They also talk about the two unique abilities of AI: Automation and Smartness, under which the hardware and software run more effectively and efficiently. They use a case study to explain how there is significant price discrimination when using a chatbot or not. According to their research, the wholesale price quotes are significantly more in case of chatbots and AI. The reverse is also true where buyers can use a smart recommendation system to get the most value, which can effectively reduce suppliers' price quote. This implies that automation isn't very valuable without smartness, which in turn suggests that building smartness is necessary before considering high levels of autonomy.

6.2.9 AI in E-Commerce and Finance – HAL Theses (Abes Star)

According to this thesis, AI has brought revolutionary changes to various sectors, including e-commerce and finance. In this thesis, they present the four applications of AI that greatly improve goods and services, enable automation, and greatly increase the efficiency of many tasks in all sectors. They start by improving the product search service by assessing a search query and then building a predictive model on daily sales to rank products according to the number of sales, in order to maximize the revenue of the company. Next, they organize the products using a classification algorithm according to the rankings and the forecasting. They then use this time-based sales prediction and the classification to predict one of the most difficult facets of commerce: **Stock**.

6.2.10   CLOUD COMPUTING WITH AI FOR BANKING AND E-COMMERCE – SPRINGER LINK

In this thesis, they researched AI in e-commerce and proposed guidelines on how information systems (IS) research could contribute to the research stream. They did a bibliometric analysis by taking bibliometric data from 4,335 documents and 229 articles. This analysis revealed that the core research themes were based on sentiment analysis, trust, personalization, and optimization. This chapter helped practitioners by providing them with an organized source of information on how AI can support their e-commerce endeavours.

6.3   RESEARCH METHODOLOGY

6.3.1   GENERAL GOAL

The goal of this chapter is to find out the impact of AI in revolutionizing the industries.

We are also trying to find what problems can be solved in commerce with the help of AI.

6.3.2   RESEARCH TYPE

We will be using the secondary data with the help of the previous published articles and charts.

We will be taking a theoretical approach by using the qualitative point of view.

6.3.3   POPULATION AND SAMPLE

The population of the proposed research is taken from the previous articles and statistics available on the web related to the impact of AI in industrialization.

6.3.4   METHOD AND TECHNIQUES

We will be reviewing the previous articles related to our topic and drawing our analysis on the basis of it.

We will then compare the data from before the implementation of AI in industries and after it was implemented, with the help of graphs.

We will then talk about the current problems that the industries face and how they can be solved with the help of AI.

6.4   ANALYSIS AND INTERPRETATION

6.4.1   GLOBAL AI IN ENTERPRISE RESOURCE PLANNING APPLICATIONS

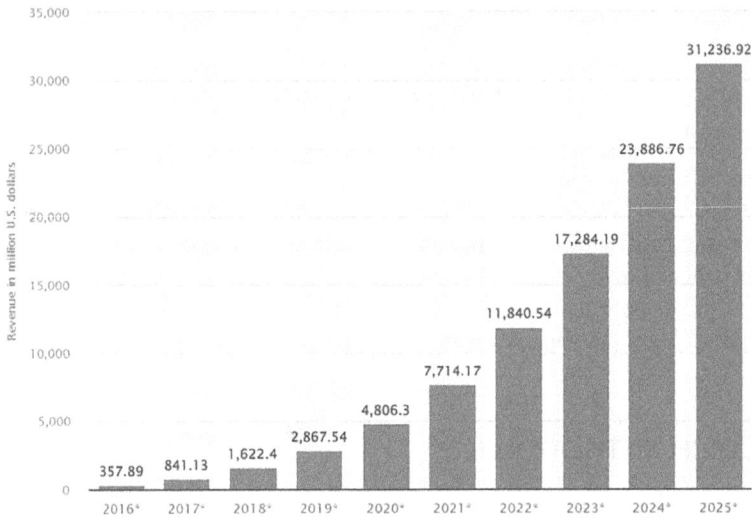

Source: Statista.

We can see exponential growth of AI in enterprise applications globally over the period of ten years (2016–2025). We also see that there is a significant change in the total revenue generated when compared to the resource planning with or without AI. This shows the importance and impact of AI in effective resource planning.

Companies are opting for smart planning to preserve precious resources. The same can be seen from the graph (4.1).

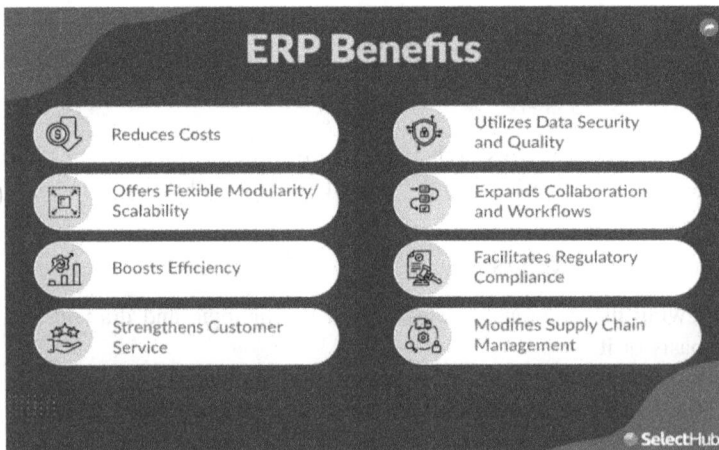

Source: Select Hub.

6.4.2 AI in Cybersecurity Market, by Region (USD Billion)

AI IN CYBERSECURITY MARKET, BY REGION (USD BILLION)

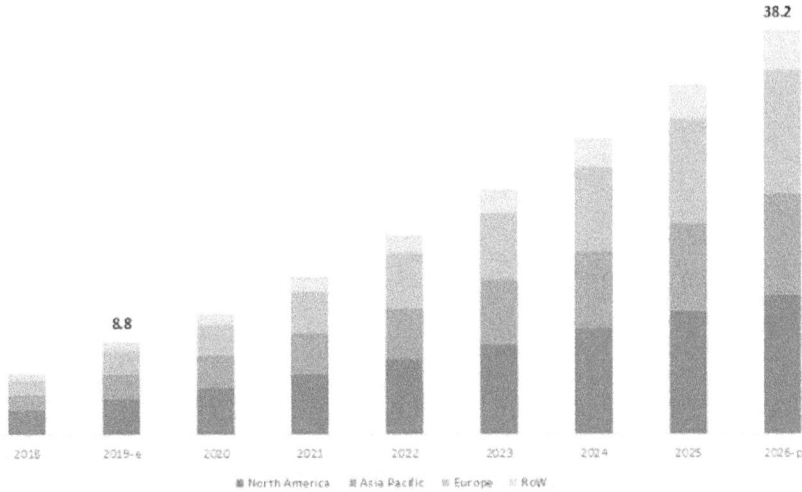

38.2

8.8

2018 2019-e 2020 2021 2022 2023 2024 2025 2026-p

■ North America   ■ Asia Pacific   ■ Europe   ■ RoW

Source: Markets and Markets.

We can see a linear growth in the graph of AI in the Cybersecurity Market for North America and Europe, whereas exponential growth in the Asia Pacific region. This may be because North America contains more developed countries, e.g., the USA, whereas the Asia Pacific region is made up of developing countries, e.g., India and China, so North America adopted the latest technologies a bit earlier compared to the Asia Pacific.

Companies in Europe and North America have very strict cyber policies because of privacy laws; thus, they have to use smart and effective AI technologies to prevent data breaches. That is why AI plays a very important role in cybersecurity. The same can be seen from the graph (4.2).

Benefits of Cybersecurity

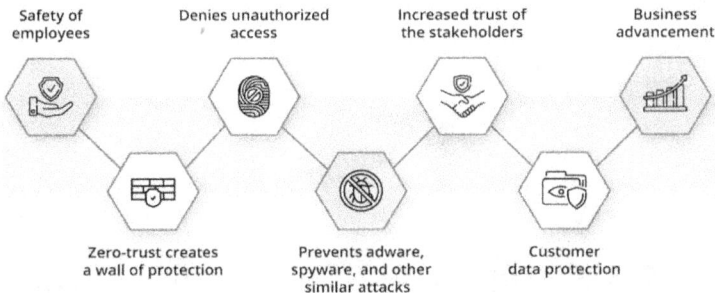

| Safety of employees | Denies unauthorized access | Increased trust of the stakeholders | Business advancement |

| Zero-trust creates a wall of protection | Prevents adware, spyware, and other similar attacks | Customer data protection |

Source: Payatu.

6.4.3   AI in Inventory Management Market Globally, from 2012 to 2025 (USD Billion)

Size of the warehouse automation market   N-iX
worldwide from 2012 to 2025 (in billion USD)
Source: Statista

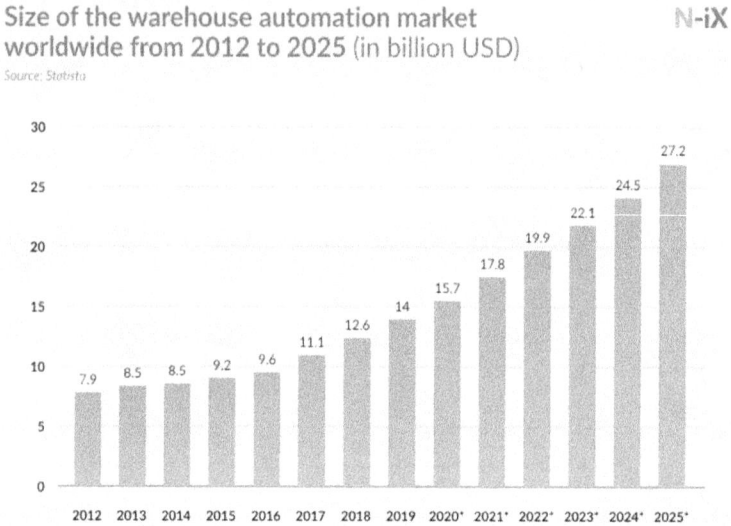

Source: Statista.

We can see a gradual shift in the graph for AI in the Inventory Management market. This shows how the integration of AI in Inventory not only makes the inventory system more efficient but also generates more revenue because of the cost reduction in manual labour. Companies are starting to implement AI in inventory management systems to solve complex tasks and reduce workloads. This system also reduces the risk of human error. The same can be seen from the graph (4.3).

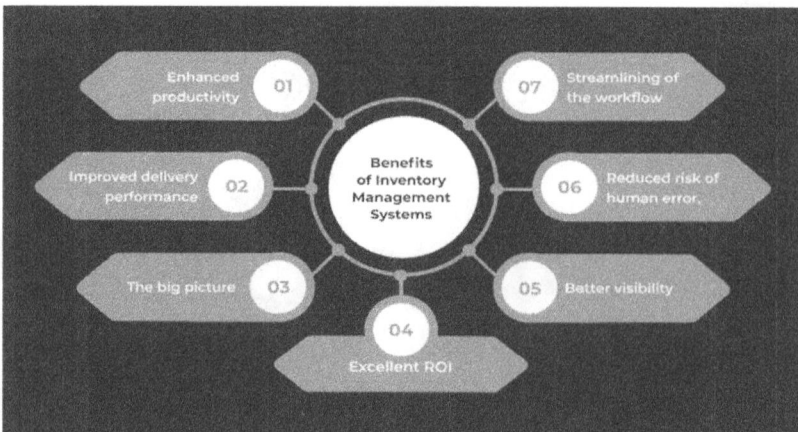

Source: Altamira.

6.4.4 Impact of AI in Supply Chain and Logistics

Impact of AI on Retailer's
Supply Chain and Logistics

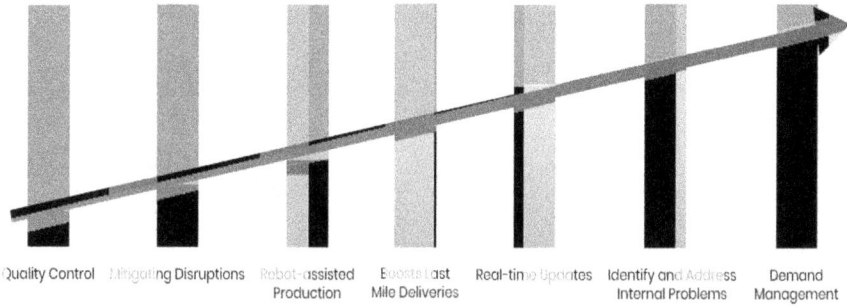

| Quality Control | Mitigating Disruptions | Robot-assisted Production | Boosts Last Mile Deliveries | Real-time Updates | Identify and Address Internal Problems | Demand Management |

Source: Oodles Technologies.

We can see that there is a very significant impact of AI on supply chain and logistics, especially in demand management, identifying problems, and finding real-time updates. AI can help in quality control, minimizing disruptions, and robot-assisted production. Companies are starting to use supply chain management systems, which are powered by AI, to improve customer experience by providing them with real-time updates and timely deliveries. They also manage demand by using sales forecasting with the help of smart AI. This not only reduces wastage of goods and services but also prevents shortage of goods. The same can be seen from the graph (4.4).

Source: Solution Dot.

6.4.5   AI in the Manufacturing Market, by Region (USD Billion)

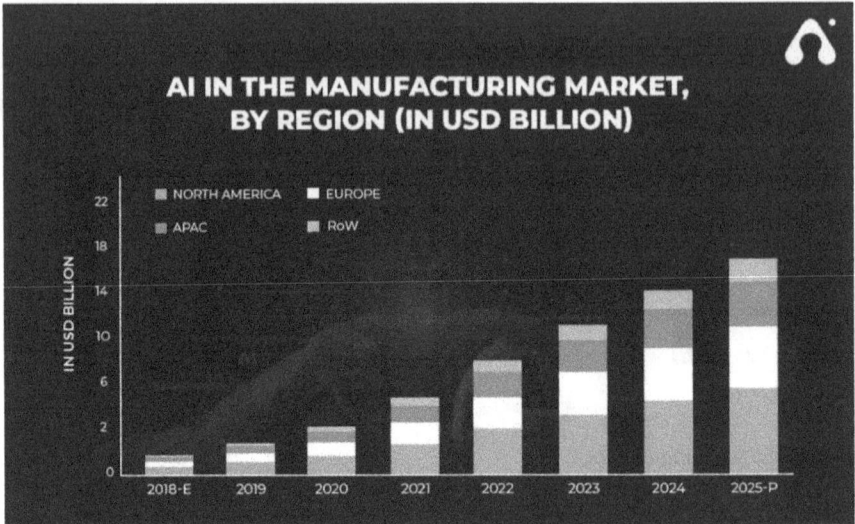

Source: Appventurez.

We can see exponential growth of AI in the manufacturing market. This graph not only shows the dependency of effective manufacturing on AI but also shows how significantly it can increase the revenue of a company by making decisions by themselves as well as finishing repetitive tasks set by the machine learning algorithm. These algorithms also learn over time thus handling the assigned processes better. These AI algorithms also don't need any breaks and aren't susceptible to human errors. Companies are using smart manufacturing systems to increase production while maintaining the overall quality of the product. These systems are time efficient, thus preventing costly delays and hence increasing the overall revenue.

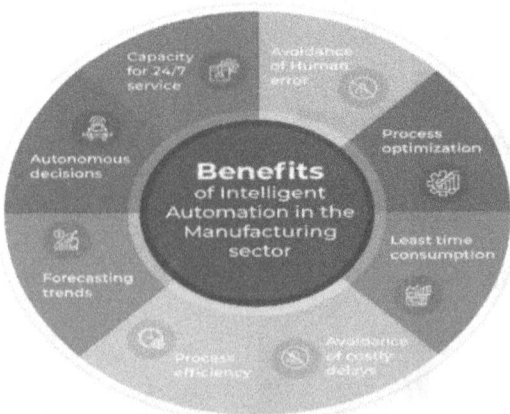

Source: 10xDS.

Attractive Opportunities in Artificial Intelligence in Manufacturing Market

Source: Markets & markets.

It is estimated that AI in the manufacturing market is expected to grow by 16.3 USD billion by 2027, and this is because of the availability of Big Data and automation technologies. The usage of these technologies enhances the operational efficiency of manufacturing units, which in turn enhances the productivity.

6.4.6 AI IN BANKING INDUSTRY WORLDWIDE, FROM 2018 TO 2030 (USD BILLIONS)

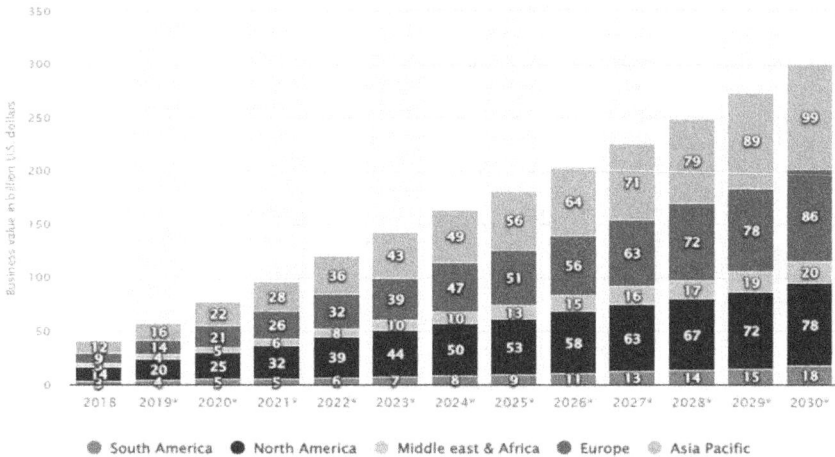

Source: Statista.

We can see a significant growth of AI in the banking industry. The involvement of AI in the banking sector reduces the complexity of the problems, e.g., investment evaluation, fraud detection. Companies are also investing in AI technologies because of their low operational costs and risks. AI can also calculate the best loan and facility evaluation that is comfortable for both the customer and the banker.

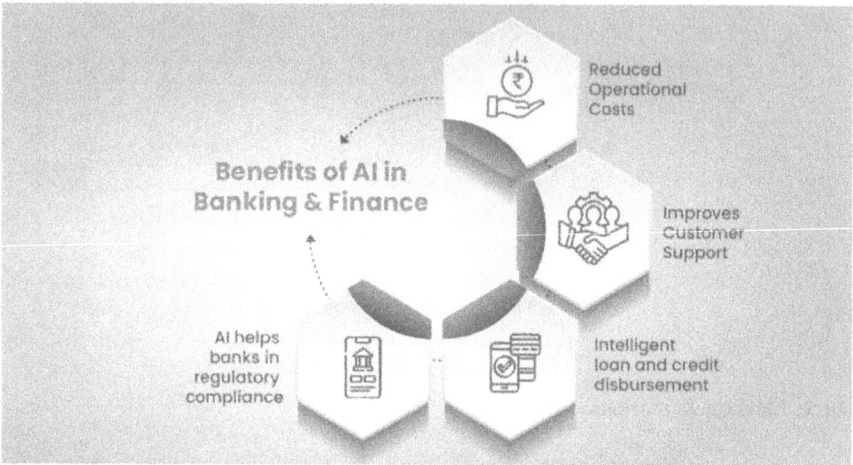

Source: ESD.

Introduction to Industrial AI: ABCDE

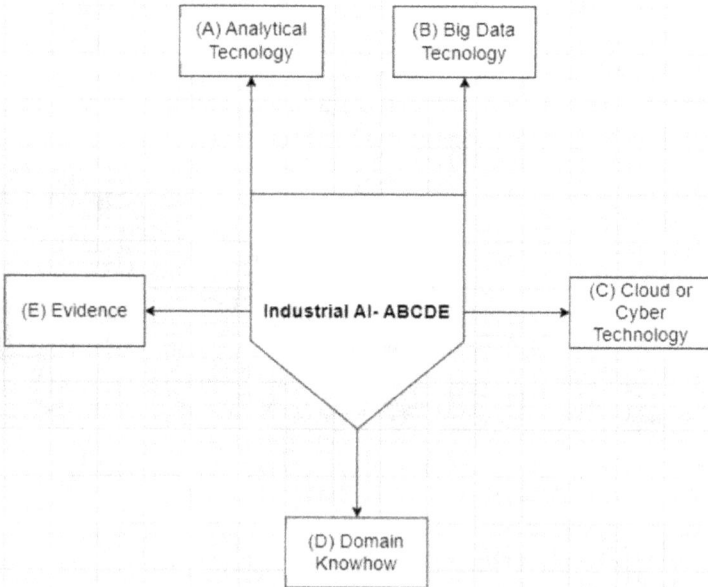

A. Analytical Technology – It is a system designed for analyzing and con-trolling the manufacturing processes, i.e., measuring quality parameters and performance attributes.

B. Big Data – The definition of Big Data is data that contain greater variety, arriving in increasing volumes and with more velocity.

C. Cloud Technology – It gives users access to storage, files, software, and servers through their Internet-connected devices.

D. Domain know-how – It is the knowledge of a specific, specialized dis-cipline or field, in contrast to general (or domain-independent) knowledge.

E. Evidence – It consists of both information that supports and corroborates management's assertions regarding the financial statements.

6.4.7 COMPARISON OF INDUSTRIAL AI TO OTHER SYSTEMS

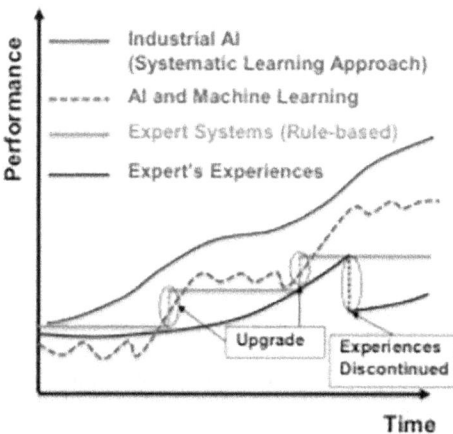

Source: Science Direct.

We can see a very important comparison between industrial AI and other learning systems. It shows how AI can overtake other systems in performance and efficiency.

Expert Systems – Expert Experience systems are based on the extent of the knowledge of the expert. The extent of their knowledge cannot cope with the latest technologies. That is why these systems are outdated. They also have a lower output and take a lot more time to solve complex problems. The graph dips when these outdated systems can't handle the complexity of the problems.

Expert Systems (Rule-based) – These systems are a bit better than the expert experience systems since they work on a set of defined rules that make autono-mous decisions on the basis of these protocols. These systems also need to be updated regularly to deal with the increasing complexity of problems; that is why the graph dips.

Industrial AI – These systems have machine learning algorithms that help the system to learn from its experiences; that is why they can keep up with the latest trends. They also have the ability to make autonomous decisions, which helps them to predict problems before they surface.

The impact of Industrial AI – from solving visible problems to avoiding invisible ones.

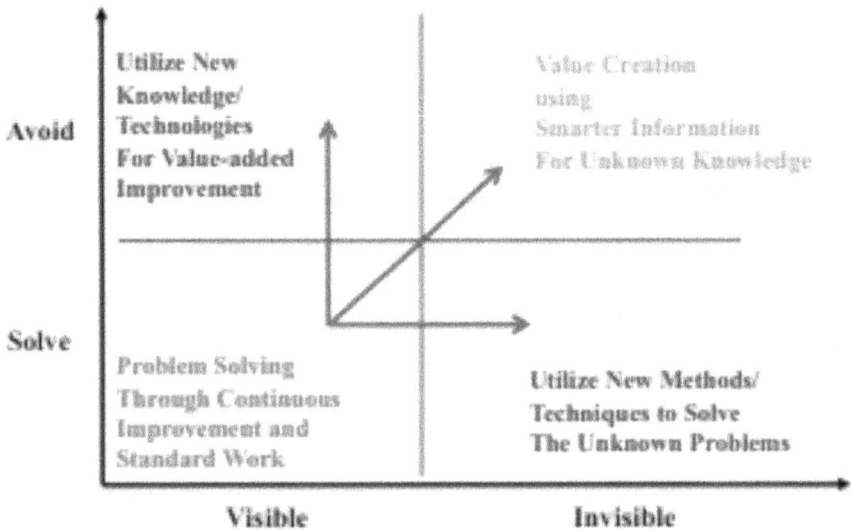

Avoid

Utilize New Knowledge/ Technologies For Value-added Improvement

Value Creation using Smarter Information For Unknown Knowledge

Solve

Problem Solving Through Continuous Improvement and Standard Work

Utilize New Methods/ Techniques to Solve The Unknown Problems

Visible   **Invisible**

Source: Science Direct.

This graph shows how AI can drive us from visible space to invisible, and from solving problems to avoiding them before they surface.

6.5   CONCLUSION

"Anything that could give rise to smarter-than-human intelligence – in the form of Artificial Intelligence, brain-computer interfaces, or neuroscience-based human intelligence enhancement – wins hands down beyond contest as doing the most to change the world. Nothing else is even in the same league." – Eliezer Yudkowsky, American computer scientist.

We can already see from the analysis and the findings how big the sectors with AI are. They are significantly reducing the labour costs and complex manual tasks. They also increase the efficiency and effectiveness of the production. It can manage all the sectors of any industry from primary to banking sectors.

BENEFITS OF
ARTIFICIAL INTELLIGENCE

AI

01 No Human Error

02 24*7 Availability

03 Unbiased Decisions

04 Quicker Decision-Making

05 No risks

06 Healthcare Applications

07 Managing Recurring Tasks

Source: Science Direct.

REFERENCES

1. Ashok. (2018). AI in supply & procurement. *IEEE*, 10.1109/AICAI.2019.8701357
2. Aria, M., & Cuccurullo, C. (2017). Bibliometrix: An R-tool for comprehensive science mapping analysis. *Journal of Informetrics*, 11(4), 959–975. 10.1016/j.joi.2017.08.007
3. Axel, Falk Andreas. (2018). Digital transformation trends: Industry 4.0, automation, and AI. *International Symposium on Leveraging Applications of Formal Methods*.
4. Bauer, J., & Jannach, D. (2018). Optimal pricing in E-commerce based on sparse and noisy data. *Decision Support Systems*, 106, 53–63. 10.1016/j.dss.2017.12.002
5. Bawack, R. E., Wamba, S. F., & Carillo, K. (2021). A framework for understanding artifcial intelligence research: Insights from practice. *Journal of Enterprise Information Management*, 34(2), 645–678. 10.1108/JEIM-07-2020-0284
6. Blöcher, K., & Alt, R. (2021). AI and robotics in the European restaurant sector: Assessing potentials for process innovation in a highcontact service industry. *Electronic Markets*, 31(3), 529–551. 10.1007/s12525-020-00443-2

7. Cao, Y., & Li, Y. (2007). An intelligent fuzzy-based recommendation system for consumer electronic products. *Expert Systems with Applications*, 33(1), 230–240. 10.1016/j.eswa.2006.04.012

8. Carbó, J., Molina, J. M., & Dávila, J. (2007). Avoiding malicious agents in E-commerce using fuzzy recommendations. *Journal of Organizational Computing and Electronic Commerce*, 17(2), 101–117. 10.1080/10919390701293972

9. Cardoso, G., & Gomide, F. (2007). Newspaper demand prediction and replacement model based on fuzzy clustering and rules. *Information Sciences*, 177(21), 4799–4809. 10.1016/j.ins.2007.05.009

10. Chang, R. M., Kaufman, R. J., & Kwon, Y. (2014). Understanding the paradigm shift to computational social science in the presence of big data. *Decision Support Systems*, 63, 67–80. 10.1016/j.dss.2013.08.008

11. Da'u, A., Salim, N., Rabiu, I., & Osman, A. (2020). Recommendation system exploiting aspect-based opinion mining with deep learning method. *Information Sciences*, 512, 1279–1292. 10.1016/j.ins.2019.10.038

12. Deniz, M. (2021). Artificial intelligence-based inventory management: A Monte Carlo tree search approach. Annals of Operations Research, 308, 415–439.

13. Digital Transformation AI & ERP (2019). Sadat Academy for Management Sciences.

14. E-commerce worldwide - statistics & facts.

15. Iqbal Md Hasan, R. (2021). AI-driven cybersecurity: An overview, security intelligence modeling and research directions. SN computer science.

16. Kar, R., & Haldar, R. (2016). Applying chatbots to the internet of things: Opportunities and architectural elements. *Inter. Journal of Advanced Computer Science and Applications*, 7(11), 147–154.

17. Market-Reports in artificial-intelligence-manufacturing-market-72679105.

18. Mohammad, S. M. Artificial Intelligence in Information Technology (June 11, 2020). Available at SSRN: https://ssrn.com/abstract=3625444 or http://dx.doi.org/10.2139/ssrn.3625444

19. Pan, Y. H., 2016. Heading toward artificial intelligence 2.0. *Engineering*, 2(4), 409–413. 10.1016/J.ENG.2016.04.018

20. Sabine, C. (2022). Development and application of a human-centric co-creation design method for AI-enabled systems in manufacturing. *IFAC Papers Online*, 55(2), 516–521.

21. Sabah Wai, C. (2019). Special issue on artificial intelligence in cloud computing. *Springer Computing*, 105, 507–511. 10.1007/s00607-021-00985-z

22. Trappey, A. J. C., & Trappey, C. V. (2004). Global content management services for product providers and purchasers. *Computers in Industry*, 53, 39–58.

23. Thiraviyam, T. (April 2018). Artificial intelligence marketing. *International Journal of Recent Research Aspects. Special Issue: Conscientious Computing Technologies*, 449–452.

24. Umamaheswaran, Ganjeizadeh, Ghasib. (2019). Inventory management and cost reduction of supply chain processes using AI based time-series forecasting and ANN modelling. *Procedia Manufacturing*, 38 (2019) 256–263.

25. Vidosav, Slavenko, Bojan, Ugljesa. (2020). ERP in Industry 4.0 context. *EIFIPAICT*, 591.

26. Vishal Dineshkumar Soni. (2020) Emerging roles of artificial intelligence in ecommerce. *International Journal of Trend in Scientific Research and Development (IJTSRD)*, -ISSN: 2456 – 6470.

27. Yang. (2017). Industry 4.0: A survey on technologies, applications and open research issues. *Journal of Industrial Information Integration*, 10.1016/j.jii.2017.04.005

28. Zbigniew Tadeusz, Ryszard, Jaroslaw, Konrad. (2019). Data-driven machine learning system for optimization of processes supporting the distribution of goods and services: A case study, 1st International Conference on Optimization-Driven Architectural Design (OPTARCH 2019)

7 Safepad

Reusable Sanitary Pads – Managing the Dynamics of Technology in Modern Day Society: The Innovating Firm in a Societal Context

Jashim Uddin Ahmed
Professor, Department of Management, North South University, Bashundhara, Dhaka, Bangladesh

Asma Ahmad
Research Associate, InterResearch, Bashundhara, Dhaka, Bangladesh

M. Jasim Uddin
Senior Lecturer, Christ Church Business School, Canterbury Christ Church University, Canterbury, United Kingdom

Quazi Tafsirul Islam
Senior Lecturer, North South University, Bashundhara, Dhaka, Bangladesh

7.1 INTRODUCTION

Women and young girls must be using the right menstrual products that are good for their health. It is also important to switch to a more sustainable and environmentally friendly option as these reusable menstrual products are a much healthier option as well. Bangladeshi women and young girls still face many social stigmas regarding menstruation, and Youth's Voice aims to spread knowledge and awareness about menstruation. For the betterment of society, it has launched the Safepad Project, which aims to provide a sustainable and healthier option for menstrual products. This campaign aims to reduce the stigma associated with menstruation.

DOI: 10.1201/9781003365525-7

WASH model of menstrual health hygiene is an approach to the prevention, promotion, and elimination of menstrual hygiene product-related diseases. This model addresses women's social, cultural, and economic rights at multiple levels. It focuses on empowerment of individuals to take control over their personal lives and the reduction of inequalities between the sexes in society. They are focusing on the need for appropriate management of menstruation in women. The WASH model includes four components: communication, provisioning/production/use, sanitation, and servicing/service delivery.

To provide high-quality care to avert and combat infection in health systems, services for WASH must be available. Universal, affordable, and sustainable access to WASH is a key public health issue within international development and is the focus of the SDGs. The menstrual hygiene management (MHM) pathway to help attain SDG 3.7 (universal access to sexual and reproductive healthcare services) and SDG 6.2 (access to adequate and equitable sanitation and hygiene) was proposed by the United Nations.

Good menstrual hygiene is significant. WASH is a crucial component of health. What is considered a basic human right, and what is not varied around the globe. Whether the government can afford to provide its inhabitants with the basic essential needs of amenities, services, and products for free is also a concern (Ahmed et al., 2022).

The use of pads is a public health win in a nation where women have historically used cloth scraps to absorb menstrual blood. However, sanitary pads present an environmental waste problem that has prompted a growing movement in favour of alternative products such as reusable menstrual pads (D'souza, 2022). Sanitary napkins and other menstruation products produce trash that affects the environment, human health, and a community's well-being once discarded. This is shown to be unsustainable now (Murthy, 2015). Nowadays, most companies are producing menstrual pads in the market, but instead of getting alleviated, the problem of environmental pollution and carbon footprints is on the rise as these are mostly made of plastic. Conventional menstrual pads are made up of three main layers having their specific functional properties: the top sheet,[1] absorbent core, and barrier sheet (Balda et al., 2021). In her lifetime, the average woman is said to use more than 11,000 feminine hygiene products – just one woman, that is. It is undoubtedly a significant waste of time and cash (Tiku, 2020). Every day, sanitary napkins are flushed down toilets or wind up as street trash. Given the overflowing landfills, disposing of solid waste is a significant difficulty for many local administrations. Pads stay in the same condition in a landfill for a very long time (Murthy, 2015).

A disposable uses throwaway plastics, which take around 700 years to decompose. This is a significant threat to the environment because disposable sanitary napkins produce about 200,000 tonnes of imperishable waste around the globe in a year (UNDP, 2022).

7.2 REUSABLE MENSTRUAL HYGIENE PRODUCTS

In underdeveloped nations, unhygienic reusable or temporary pads are still used to collect menstrual blood. Women in these nations frequently revert to using old fabric, old mattress foam, or even infection-causing materials like leaves, husks, discarded cement bags, etc., when they are menstruating. Women and teenage girls all across the world suffer from a lack of access to feminine hygiene products. Depending on the product's availability and price, girls and women use several different products on the market to regulate their monthly bleeding. Since many women are still unable to even get the items

TABLE 7.1

Reusable Menstrual Products on the Market

Type	Strengths	Weaknesses	Cost (USD)
Cloth menstrual pads	• Reusable, eco-friendly • Breathable and convenient to use • Menstrual blood noticed easily when washing	• Time-consuming cleaning method is troublesome and unhygienic • Blood overflow if not changed timely	USD 10/piece
Menstrual cups	• Reusable, eco-friendly • No irritation when worn, convenient to use • Blood overflow is not a matter of concern • Menstrual blood noticed easily when washing	• Time-consuming cleaning method is troublesome and unhygienic • Prone to infection due to prolonged usage but less frequent than tampons • Need to understand how to use and adjust to it	USD 20/piece
Menstrual underpants	• Reusable, eco-friendly • No irritation when worn, convenient to use • Blood overflow is not a matter of concern • Does not need to adapt to new changes	• Time-consuming cleaning method is troublesome and unhygienic • Requires alternative sanitary products during heavy flow	USD 20/piece

Source: Tu et al. (2021); p. 5 (modified and reduced contents).

in rural areas, the safety and dignity of utilizing these products are secondary (Pednekar et al., 2022). Rural women use reusable clothes, but they have a limited capacity to absorb moisture for usual usage durations and are seen as unsanitary (Achuthan et al., 2021).

Most menstrual hygiene products manufactured worldwide are disposable, and the remainder can be reused. "Cloth pads, menstrual cups, and menstrual underwear are the top trends among women's views toward environmentally friendly menstruation goods, although menstrual pads and tampons make up the majority of menstrual product sales" (Tu et al., 2021). There are three different types of reusable hygiene products: reusable and washable fabric pad, menstrual cups, and menstrual undergarments (as shown in Table 7.1).

7.3 REUSABLE AND WASHABLE CLOTH PADS

Soft and breathable fabrics are used to make cloth pads, promoting airflow. This implies that when women use reusable cloth pads, there will be less burning, discomfort, and stinging. Wearing cloth pads reduces the risk of developing a yeast infection since some women are susceptible to the bleaching chemicals used to achieve the pure white appearance of disposable pads. Changing to cloth pads is a huge relief for many people (Tiku, 2020). However, these reusable sanitary clothes need to be cleaned very carefully and dried in the sun to maintain hygiene, even though these reusable pads are environmentally friendly. The sun is a natural sterilizer, thus drying the cloth pad in the sun sterilizes it for later use. The reusable cloth pads are economical, conveniently accessible, and sustainable. The cloth pads need to be stored in clean storage to prevent any sort of contamination (Kaur et al., 2018).

7.4 MENSTRUAL CUPS

Menstrual cups are a reusable product with the potential to improve menstrual management among poverty-period women and girls in developing countries (Madziyire et al., 2018). Menstrual cups are an innovation for underprivileged women and a great alternative to tampons and sanitary pads. These cups resemble silicon rubber cups, which makes them simple to fold and put inside the vagina for menstrual blood collection. Depending on the quantity of menstrual flow, they can be worn for up to 6–12 hours, requiring less frequent removal and cleaning. They are safe for the environment and reusable. It provides a practical, affordable, and sustainable option for poor sanitation conditions (Kaur et al., 2018).

7.5 MENSTRUAL UNDERPANTS

Menstrual underwear is made up of three layers: an absorbent material that can contain the equivalent of one to two tampons' worth of flow; a moisture barrier to keep the user comfortable; and a layer intended to stop any leaks or stains (Grove Collaborative, 2022). Since a lot of bacteria exist between clothing and human skin and are hazardous to people, underwear clothing must include antibacterial characteristics (Shanmugasundaram and Gowda, 2011).

7.6 SAFEPAD: AN OVERVIEW

7.6.1 BACKGROUND

In 2011, Youth's Voice was launched by a college student, Tahmid Kamal Chowdhury, who is also the CEO of Safepad Bangladesh. Youth's Voice is the youth branch of Youth Worldwide Foundation (YWF), a non-profit organization that promotes education in Bangladesh's rural areas. Tahmid and his colleagues concluded that experience that Bangladeshi culture, particularly in rural regions, severely disregards MHM. This encouraged him to put this on the YWF agenda. In addition to launching MHM programmes in Bangladeshi schools, YWF began providing disposable sanitary products in 2015 as well.

7.6.2 INITIATIVES AND OBJECTIVES

In Bangladesh, 50% of adolescent girls and 64% of adult women do not have access to proper sanitary napkins during menstruation due to poverty and social taboos. They use unclean and unsanitary old rags during the menstrual cycle, which results in illness, absence from school, and gender and social injustice (Cordaid, 2022). The purpose of this initiative is to end the taboo around menstruation, dispel common beliefs and misconceptions, educate boys and girls about menstruation, and provide sanitary products to underprivileged girls who are menstruating. The project wants to broaden its reach into underprivileged areas and provide teenagers with the skills they need to understand and manage menstrual safety. By using safe, hygienic, and reusable sanitary pads, Safepad hopes to increase the percentage of young girls managing their menstrual hygiene in a sustainable, environmentally friendly, and correct manner. Safepad, an anti-microbial fabric-made reusable sanitary napkin, has been distributed in 37 districts for adolescent girls and women starting in 2019. In addition to pads, the team has been trying to increase knowledge of menstrual hygiene management among teenagers and their parents.

7.6.3 PRODUCT AND PROCESS

Reusable Safepad sanitary pads protect against the danger of infection associated with reusable sanitary pads due to a permanently bonded anti-microbial technology that eliminates bacteria, fungus, and other disease-causing microorganisms and also enables the sanitary napkins to have maximum absorption capacity (The Business Standard, 2022) (Figure 7.1).

The pad is constructed of three layers: a top layer of fuchsia microfiber fabric; three layers of white microfiber fabric in the centre; and a polyurethane laminate anti-leak security layer at the base. Microfiber fabric makes up the outer layer, which is the

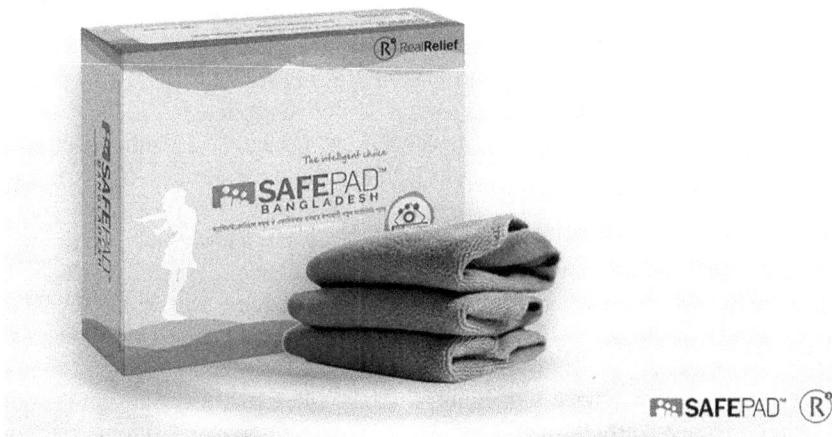

FIGURE 7.1 Safepad product.

Source: Safepad Bangladesh Facebook page, 2020.

area in touch with the genitalia. The softness of this fabric makes it pleasant to wear and conveniently clean, which prevents traces of blood from adhering to it. White microfiber fabric makes up the core portion. There is no additional odour or water-soluble colouring in this layer. A layer of polyurethane laminate has been added at the bottom to stop the blood from spilling out. This was challenging since laminate may melt at high temperatures. However, since it is a reusable napkin, the users mostly like going to wash it with warm water.

The process for producing the pads includes four steps – laser cutting, assembling, quality check, and packing. These are done in three different factories.

The fabric that is used to construct the pads is imported from India. After producing and assembling the product, it is sent to the headquarters where quality control check is carried out. When all the quality requirements are fulfilled, the products are sent to the inventory and distributed to different locations to raise awareness among the population (The Business Standard, 2022).

7.7   SAFEPAD SIZE AND PRICE

For days with strong flows and days with regular flows, Safepad provides two sizes. The pads come in a colourful recycled fuchsia paper package. The regular pack has four pads (one heavy flow and three normal flows), which costs BDT 360, whereas the economy pack only contains two pads (one heavy flow and one normal flow), costing BDT 190. The price of the Safepad may appear to be somewhat greater than that of other brands that sell 10 disposable pads in packets costing BDT 170–180.

7.8   COMMUNITY ENGAGEMENT

In addition to offering reusable pads, Safepad has reached out to Rohingya women[2] and has provided a means of employment so that women like Tayeba may learn how to produce and sell pads in their neighbourhood. Only women are employed to make these pads in their official factory in Chattogram.[3] Additionally, they have donated sewing machines for the Rohingya women to learn how to create the pads in their skill development centre in the Balukhali Rohingya camp in Cox's Bazar,[4] Bangladesh. There are now 54 women learning to create pads in two sessions.

Furthermore, Real Relief partnered with Forum for Public Health to set up a Safepad production centre in Kutupalong Rohingya refugee camp in Cox's Bazar[5] as a part of the UNHCR Self Reliance Project Activities. The production centre seeks to improve the livelihood of the Rohingya women and to make them independent so that they are able to attain their civil, cultural, economic, and social rights. This project aims to amplify women's empowerment and gender equality while at the same time contributing to long-term economic growth and societal development. In 2020, the refugee women sewed, finished, and packaged the first 128,000 Safepads. The success of Safepad's reusable sanitary napkin production will help mitigate poverty through sustainable production practices and the creation of jobs for underprivileged women (Real Relief, 2020).

Moreover, during the devastating flash flood of 2022, where an estimated 7.2 million people were severely affected, Safepad Bangladesh reached out to the flood-affected women in Sylhet with reusable sanitary napkins they could use during the difficult times (Safepad Bangladesh, 2022).

Additionally, in 2019, Safepad participated in the "Menstrual Health Management Walk" organized by Youth Worldwide Foundation to raise awareness about menstruation and break the taboo. A vast number of participants gathered, most of whom were females. One participant mentioned the campaign being "Impactful along with Successful" as the passersby were curious about the purpose of such a huge assembly (Safepad Bangladesh, 2019).

On top of that, in 2021, Safepad Bangladesh participated in the virtual transnational "Kick Off the Dialogue Campaign" to celebrate International Menstrual Health Day on May 28. The aim of this campaign was to inform people about menstrual hygiene and health and tackle stigmas surrounding these topics. Real Relief, Global Goals World Cup, UNYA Denmark, and Safepad joined together to discuss some of the challenges related to menstrual health and hygiene in different countries. The countries together "symbolically kicked off the Sustainable Development Goal (SDG) football from country to country" to give a glimpse of the SDGs that the individual countries have met by providing reusable sanitary napkins to the female population of their respective countries (Safepad Bangladesh, 2021).

The 4A framework consists of four key concepts: Affordable, Available, Awareness, and Acceptability. Affordability is affordability in terms of price paid. Availability is whether or not the product can be purchased by a population. Awareness is why someone buys the product and its acceptance by buyers. According to Ladefoged and Hargadon, "Affordability encompasses issues such as affordability at multiple price points and perceptions about the willingness to pay for alternative products."

7.9   4A'S MENSTRUAL HYGIENE

There is a sizable market for goods and services provided to the world's underprivileged. More than 4 billion people are consumers at the very bottom of the economic pyramid (BOP), defined as those with per capita incomes of less than $1,500. The creation of a strategy that meets the 4As – availability, affordability, acceptability, and awareness – is at the core of almost all of these organizations' success (Anderson and Billou, 2007). The 4As model of service delivery (as shown in Figure 7.2) provides a useful way of thinking about the issues involved (Ahmed et al., 2019).

7.9.1   AVAILABILITY

Due to a lack of access to appropriate information and sufficient understanding, the majority of women and girls frequently have difficulties with menstrual hygiene (Dutta et al., 2016). There are many taboos regarding menstruation in society. Women and girls face many problems regarding this. The world has become much more progressive in many matters. However, in third-world countries, women and girls still feel ashamed to carry and purchase their menstrual products openly.

Disposable sanitary pads are available almost everywhere, as they are considered essential for women. It's easily available even in neighbourhood retail stores or pharmacies. However, this is the case in urban areas, not in rural areas. As these

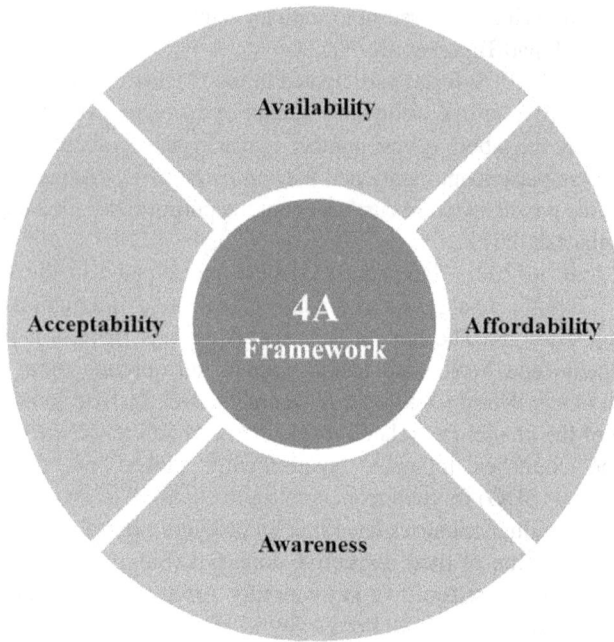

FIGURE 7.2   The 4A framework.

Source: Authors' own creation based on Anderson and Billou (2007).

sanitary pads are quite expensive for rural women, they don't buy them very often, so there is no market for these pads there, leading to less availability.

The concept of using reusable sanitary pads is still very new in Bangladesh, whereas in western countries people have started to understand the importance of reusable pads, and many are shifting to menstrual cups as they are reusable. Thus, reusable sanitary napkins are not easily available in any store. Many shops may not want to keep them in there as they know women may not buy them, and advertising them on the spot to the customer may not be comfortable. This is due to the lack of awareness about the importance of sustainability.

7.9.2   AFFORDABILITY

The low socioeconomic position makes it more difficult to get period care items at reasonable prices (Madziyire et al., 2018). Disposable sanitary napkins are quite hard for many lower-income households to afford. The foreign brands cost quite a lot, ranging from BDT 300–500, while the local brands cost a minimum of BDT 100. The government also imposes tariffs on these sanitary napkins, leading to an increased price and making them more expensive. According to World Bank national accounts data, in Bangladesh the per capita income over has risen to USD 2,503 (World Bank, 2022); this progress has not been diffused among the entire population (Ahmed et al., 2019). It is hard for women in rural areas to afford these disposable sanitary napkins as well, so they reuse clothes, which aren't suitable and are kept in an unhygienic manner. The reusable sanitary napkins, such as Safepad Project, are a bit expensive for 4–6 packs in each. However, these are reusable and hygienic, thus buying them once means women and girls can use them quite a few times. Using reusable pads can save money each month, rather than buying sanitary pads each month, which can total BDT 500. Buying reusable sanitary products once, ranging from BDT 190–360, can be used for a few months as a period.

7.9.3 ACCEPTABILITY

Women are using disposable sanitary napkins quite a lot, especially women in urban areas. As it's disposable, there is no hassle in washing; it's more hygienic. The concept of reusable sanitary napkins is still quite new, as they have to be washed, then carefully dried, and stored in a very clean place. Women who use a disposable sanitary pad may be very skeptical about using these reusable sanitary napkins, as they are uncertain about how much these reusables may restrict blood flow. Many may consider reuse very unhygienic.

7.9.4 AWARENESS

Indeed, there is a lack of awareness about the importance of sustainability. Taking care of health is crucial. Disposable sanitary napkins may be easier and more comfortable to use, but they do have consequences for health and the environment. Disposable sanitary napkins may contain heavy blood flow, but these pads have chemicals that may harm the woman in the long term. Wearing these pads for a long time may also cause many health issues. Disposing of plastic has been a huge concern and has been quite difficult. Changing to reusable pads will have a better effect on the environment and health. Women and young girls need to understand the importance of reusing these pads both for the environment and their health. The awareness and knowledge regarding how important it is to shift the habit of using disposable pads to reusable pads.

7.10 SETBACKS AND CHALLENGES

Women and girls face numerous difficulties caused by stigmas and taboos surrounding menstruation. They are subject to cultural constraints, which must be lifted since they affect how women efficiently manage their periods. Women have logistical difficulties when using menstrual products, such as the need for proper facilities that are female-friendly and the secrecy surrounding the purchase of sanitary goods. Women had a hard time buying menstruation products to control their periods, and women were particularly affected by the pandemic since they could not purchase or receive menstrual pads (Pednekar et al., 2022).

Different cultures have different views regarding menstruation, and these taboos are still believed and followed. From developed to developing countries, young girls skip school due to their menstruation. Part of it is due to the embarrassment they may face if anything goes wrong, and sometimes due to the unwellness caused by menstruation. In parts of the world, women and girls are expected to stay in their rooms during menstruation and not come in front of other men. Sometimes during menstruation, touching anything means it becomes impure. The notion that women are impure, unpleasant, or immoral while menstruating is among the most prevalent period taboos.

In Bangladesh, most schools lack the understanding of how to educate students regarding reproductive systems properly. The "menstruation" topic still seems to be very sensitive to talk about, so people aren't comfortable talking about this in front of boys or men. However, knowledge regarding menstruation is universal; both men and women should have adequate information.

The operation manager of Safepad, Akil Mustaseem, said that the teachers and guardians were not comfortable talking about periods and women's reproductive organs. They were told to remove any part of the presentation that shows women's reproductive systems and how menstruation works. They were asked to follow as instructed by the teacher; if not, they would be removed from presenting anything to the school.

7.11   FUTURE PERSPECTIVES

The organization's team will work together with the administrators of the targeted schools. They will work with community organizers, community health volunteers, and locally run organizations to connect with out-of-school teenagers. The project team will then concentrate on educating the users and other stakeholders and provide the participants with informational pamphlets and surveying tools. Teenage girls will also get a donation of one pack of reusable pads, which has four pieces. In addition, the project team will establish varying degrees of collaboration with the government of Bangladesh to support sustainable development.

Women's health, adequate facilities, equity, dignity, and respect can result in disregard if there is a lack of knowledge, even at the policy level. Focused information about reusable menstrual products, replacement schedules, and disposal is urgently needed. To enhance excellent MHM practice from childhood, it is essential to increase teaching and knowledge about menstrual hygiene at school even before early puberty. Sanitation, education, and sanitary facilities should all be expanded and improved with assistance from the government and private organizations. Teaching young girls about cleanliness, menstruation, sanitation, and physiology is essential to lower menstrual sickness rates (Pednekar et al., 2022). The Safepad Project offers instant responses to several issues that girls in poor environments face when they are in menstruation. The Safepad offers a long-term solution to issues including the usage of unhygienic materials and an increase in school dropout rates (Project Proposal – The Safepad Project, n.d).

To make Safepad more accessible to students, they plan to install an IoT-based vending machine in the school through which reusable pads can be easily purchased. The basic idea is to give students a card with a brash or rocket number, which they may recharge to enable them to purchase and retrieve pads from vending machines.

FUNDING

This work was supported by InterResearch, Bashundhara, Dhaka, Bangladesh (2021).

DISCLOSURE STATEMENT

No potential conflict of interest was reported by the author(s).

NOTES

1 Top sheet frequently transfers the fluid to the inner layers. Next to the top layer, the absorbent core is made up of SAP and wood pulp. Barrier sheet is fluid impermeable film made up of polyethylene (Woeller and Hochwalt, 2015).
2 Rohingya women are females of the ethno-religious minority group originating in Rakhine State in Myanmar (Kojima, 2015). An estimated 745,000 Rohingyas are living as refugees in Cox's Bazar, where 52% of them are women and girls (ISCG, 2019).
3 Chattogram, an administrative zone of Chittagong division, is a district in south-eastern part of Bangladesh. It is a geographically diverse region comprised of hills, sea, valleys, and forests (Government of Chattogram, n.d.).
4 Balukhali Rohingya camp is a refugee camp situated in Cox's Bazar, Bangladesh.
5 Kutupalong Rohingya refugee camp, located in Cox's Bazar, is presently the largest refugee camp in the world. It shelters about 800,000 refugees who escaped the violent tyranny of the Myanmar government (Malteser International, n.d.).

REFERENCES

Achuthan, K., Muthupalani, S., Kolil, V. K., Bist, A., Sreesuthan, K., & Sreedevi, A. (2021). A novel banana fiber pad for menstrual hygiene in India: A feasibility and acceptability study. *BMC Women's Health*, *21*(1), 1–14.
Ahmed, J. U., Islam, Q. T., & Ahmed, A. (2022). Menstrual hygiene products as a basic need: What can Scotland's action campaign teach about improving public health worldwide? In *SAGE Business Cases*. SAGE Publications: SAGE Business Cases Originals. Retrieved January 23, 2022, from https://sk.sagepub.com/cases/menstrual-hygiene-products-basic-need-scotland-action-campaign
Ahmed, J. U., Rahanaz, M., & Rubaiyat-i-Siddique. (2019). Friendship floating hospitals: Healthcare for the riverine people of Bangladesh. *Journal of Developing Societies*, *35*(1), 175–194.
Anderson, J., & Billou, N. (2007). Serving the world's poor: Innovation at the base of the economic pyramid. *Journal of Business Strategy*, *28*(2), 14–21.
Balda, S., Sharma, A., Capalash, N., & Sharma, P. (2021). Banana fiber: A natural and sustainable bioresource for eco-friendly applications. *Clean Technologies & Environmental Policy*, *23*(5), 1389–1401.
Cordaid. (March 21, 2022). Fighting period poverty: How a sanitary pad can bring social justice (and helps save the planet). Retrieved November 02, 2022, from https://www.cordaid.org/en/news/fighting-period-poverty-how-a-sanitary-pad-can-bring-social-justice/
D'souza, N. (2022). Reusable feminine hygiene products could curb a growing waste crisis in India. Retrieved October 27, 2022, from https://fortune.com/2022/08/29/reusable-feminine-hygiene-products-could-curb-a-growing-waste-crisis-in-india/

Dutta, D., Badloe, C., Lee, H., & House, S. (2016). *Supporting the rights of girls and women through menstrual hygiene management (MHM) in the East Asia and Pacific Region: Realities, progress, and opportunities.* Bangkok: UNICEF East Asia & Pacific Regional Office (EAPRO).

Government of Chattogram. (n.d.). Chattogram at a Glance. Retrieved November 02, 2022, from http://www.chittagong.gov.bd/en/site/page/jQWu-%E0%A6%8F%E0%A6%95-%E0%A6%A8%E0%A6%9C%E0%A6%B0%E0%A7%87-%E0%A6%9A%E0%A6%9F%E0%A7%8D%E0%A6%9F%E0%A6%97%E0%A7%8D%E0%A6%B0%E0%A6%BE%E0%A6%AE

Grove Collaborative. (2022, July 13). *Beginner's Guide to Period Underwear: Everything You Need to Know About Switching.* Retrieved from https://www.grove.co/blog/period-underwear-guide

Inter-Sector Coordination Group (ISCG). (March 2019). Gender Profile No.2 for Rohingya Refugee Response. Retrieved November 02, 2022, from https://www.humanitarianresponse.info/sites/www.humanitarianresponse.info/files/assessments/iscg_gender_profile_no._2_rohingya_refugee_response_30march2019.pdf

Kaur, R., Kaur, K., & Kaur, R. (2018). Menstrual hygiene, management, and waste disposal: Practices and challenges faced by girls/women of developing countries. *Journal of Environmental & Public Health*, 1730964, 9 pages.

Kojima, Y. (December 08, 2015). *Rohingya Women in Migration: Lost Voices.* Our World. Retrieved November 02, 2022, from https://ourworld.unu.edu/en/rohingya-women-in-migration-lost-voices#:~:text=Rohingya%20women%20and%20girls%20are%20often%20referred%20to%20as%20%E2%80%9Clambs,keep%20them%20as%20their%20mistresses

Madziyire, M. G., Magure, T. M., & Madziwa, C. F. (2018). Menstrual cups as a menstrual management method for low socioeconomic status women and girls in Zimbabwe: A pilot study. *Women's Reproductive Health*, 5(1), 59–65.

Malteser International. (n.d.) Inside the Kutupalong refugee camp, Cox's Bazar. Retrieved October 10, 2022, from https://www.malteser-international.org/en/our-work/asia/bangladesh/life-in-a-refugee-camp.html#:~:text=Kutupalong%2C%20located%20in%20the%20coastal,violent%20persecution%20in%20neighboring%20Myanmar

Murthy, L. (2015). It's time to take the bull by the horns–menstrual product debris can be reduced by using Uger fabric washable pads. *Design for sustainable well-being and empowerment: Selected papers.* Bangalore: IISc Press and TU Delft, pp. 125–140.

Naher, K. (2022). *Safepad Bangladesh: Bridging the gap between women and menstrual health management.* Retrieved November 03, 2022, from https://www.tbsnews.net/features/panorama/safepad-bangladesh-bridging-gap-between-women-and-menstrual-health-management

Pednekar, S., Some, S., Rivankar, K. & Thakore, R. (2022). Enabling factors for sustainable menstrual hygiene management practices: A rapid review. *Discover Sustainability*, 3, 28.

Real Relief (2020). Safepad reusable sanitary pads empowering Rohingya refugee women in Bangladesh. Retrieved December 22, 2021, from https://www.realreliefway.com/news/safepadtm-reusable-sanitary-pads-empowering-rohingya-refugee-women-in-bangladesh

Safepad Bangladesh. (August 18, 2020). A great reusable menstrual pad should be comfortable, affordable and well made … [Image]. Facebook. Retrieved November 01, 2022, from https://www.facebook.com/safepadbangladesh/photos/pb.100063880570686.-2207520000./1651240185038633/?type=3

Safepad Bangladesh. (June 25, 2022). Emergency preparedness is a team sport. Safepad Bangladesh Team is … [Post]. Facebook. Retrieved July 23, 2022, from https://www.facebook.com/safepadbangladesh/photos/a.1206372682858721/2292046390958006/

Safepad Bangladesh. (April 30, 2021). Tomorrow we "Kick off the Dialogue Campaign" on Menstrual Health … [Video post]. Facebook. Retrieved November 03, 2022, from https://www.facebook.com/safepadbangladesh/videos/2824317814547484

Safepad Bangladesh. (November 18, 2019). This Saturday Hundreds of people joined the Menstrual Health Management Walk Marathon in … [Post]. Facebook. Retrieved November 03, 2022, from https://www.facebook.com/safepadbangladesh/videos/2312204835736801

Shanmugasundaram, O. L., & Gowda, R. M. (2011). Study of bamboo and cotton blended baby diapers. *Research Journal of Textile and Apparel, 15*(4), 37–43.

The Business Standard. (July 28, 2022). নারীদের জন্য পুনরায় ব্যবহার উপযোগী স্যানিটারি ন্যাপকিন তৈরি করছে সেইফপ্যাড বাংলাদেশ [Video]. Youtube. Retrieved November 03, 2022, from https://www.youtube.com/watch?v=d-54dgd_jiw

Tiku, S. D. (2020). Design and development of feminine reusable pad without pad holder. *International Journal of Clothing Science and Technology, 32*(2), 271–283.

Tu, J. -C., Lo, T. -Y., & Lai, Y. -T. (2021). Women's cognition and attitude toward eco-friendly menstrual products by consumer lifestyle. *International Journal of Environmental Research & Public Health, 18*(11), 5534.

United Nations Development Programme (UNDP). (August 02, 2022). Safepad Bangladesh: Bridging the gap between women and menstrual health management. Retrieved September 13, 2022, from https://www.undp.org/bangladesh/news/safepad-bangladesh-bridging-gap-between-women-and-menstrual-health-management#:~:text=Starting%20in%202019%2C%20Safepad%20Bangladesh,Menstrual%20Hygiene%20Management%20(MHM)

Woeller, K. E., & Hochwalt, A. E. (2015). Safety assessment of sanitary pads with a polymeric foam absorbent core. *Regulatory Toxicology and Pharmacology, 73*(1), 419–424.

World Bank (2022). GDP per capita – Bangladesh. Retrieved October 27, 2022, from https://data.worldbank.org/indicator/NY.GDP.PCAP.CD?end=2021&locations=BD&start=2021

8 Sustainability through Transformative Technologies

Green Banking and SDG-13

Navita Mahajan, Vibha Singh, and Navleen Kaur
Amity University, Noida, Uttar Pradesh

Oridoye Tomide Hakeem
Chi Limited, Lagos, Nigeria

8.1 SUSTAINABILITY: *GREEN BANKING*: A KEY GOAL FOR THE FUTURE

The most serious problem human beings are facing during present times is man-made environmental degradation. Along with the new challenges of globalization, there has emerged considerable interest towards the development of new ways of addressing the diverse impacts of business and society (Crane & Matten, 2010). Many of these impacts are far-reaching and profound. To mention just a few, one needs to think of impacts such as,

- Environmental pollution, in particular the effect on climate change caused by the production of transportation, the use of products such as cars, refrigerators, or newspapers.
- The ever-increasing problems of waste disposable and management as a result of excessive product packaging and the dominance of the throwaway culture.
- The erosion of local cultures and environments due to influx of mass tourism in places.

Hence, there have been continuous endeavours across the globe to measure and mitigate this problem caused by human activities. At a time of soaring energy costs and fragile financial markets, executives are increasingly looking toward sustainability strategies to save money, cut emissions, and as many claims, simply do the right thing. Faced with such problems, it has been suggested that goals and consequences of business require radical rethinking. Following the Rio Earth Summit of 1992, one concept, in particular, appears to have been widely promoted as the

DOI: 10.1201/9781003365525-8

essential new conceptual frame for assessing not only business activities especially, but industrial and social development more generally, the concept of sustainability (Crane & Matten, 2010). But sustainability can be a big part of the solution. SDG-13 includes five targets: strengthen resilience and adaptive capacity to climate-related disaster, integrate climate change measures into policies and planning, build knowledge and capacity to meet climate changes, implement the UN Framework Convention on Climate Changes, and promote the mechanism to raise the capacity for planning and management. In addition to data centre optimization and other operational issues such as teleporting, companies are now leveraging IT capabilities to facilitate sustainability initiatives across the enterprise in new areas, including fleet management, paperless billing, and construction and facility management, among others.

Banks as responsible corporate citizens are *more* concerned about social and environmental aspects because economic performance of a company is no longer considered to be the only factor for corporate success. The sustainability assessment focuses mainly on management processes rather than on actual sustainability of the company or its products. Evidence of policies and management tools feature more prominently than concrete omission data or resource consumption. It has been proved that the index does not identify sustainable companies, but rather those making process progress towards addressing the issues. Now organizations understand that if they will not focus on sustainability, they have to pay for it in the form of tangible or intangible costs. For creating a balance between businesses and the environment, green practices came into existence. Banks have started going green. Going green refers to adopting knowledge and practices that are environmentally friendly as well as ecological policies and procedures that cannot harm the environment and will sustain a business as well as natural resources for current and future generations. Some of the prominent green banking products available are,

Specialized Instruments	Environmental Rights Financing	International Green Credit Lines
Loans		
Solar Power	Environmental Credits	ADB Green Building Programme
Energy Efficiency		
Waste Management	Emission Allowance	IFC Energy Programme
Green Bonds		AFD Energy and Renewable Programme
Banking Securities	Carbon credits	KFW Green Energy Programme

Source: Climate Policy Initiative.

After the Industrial Revolution in 1991, environment degradation has been increasing day by day. The whole world is concerned about environmental issues.

There is no universally accepted definition of green banking (Alexander, 2016), and it varies widely between countries. However, some researchers and organizations tried to come up with their own definition. The Indian Institute for Development and Research in Banking Technology (IDRBT), which was established by the Reserve Bank of India, defined green banking as an umbrella term referring to practices and guidelines that make banks sustainable in economic, environmental, and social dimensions (IDRBT, 2013). These concerns are arising as green management is applying eco-friendly practices in organizations. As businesses are going globalized, organizations are shifting their structure from financial status to sustainable efficiency-based organizations. Employee development towards going green contributes a vital role in suitability of an organization. Banks and financial institutions should adopt strategies and practices that could help in reducing carbon emissions. Green banking is similar to the concept of ethical banking, which starts with the aim of protecting the environment, as it involves promoting environmental and social responsibility while providing excellent banking services (Bihari, 2011).

Green banking affects both internal and external customers of an organization and helps in reducing internal carbon footprints as well as external carbon emissions. The internal footprint can be reduced by eliminating the use of heavy lighting, air conditioning, and high paper waste and by using automation, a water consumption mechanism, and renewable energy. External carbon emission can be reduced by financing those companies and projects that are working toward pollution reduction. Banks offer various channels to access their product and service through ATMs, mobile banking, E-wallets, Internet banking, etc. Banks are comfortable in providing loans to organizations that follow the rules and regulations of environmental protection; such organizations have less chance of being a victim of natural calamities and bad goodwill risk.

The banks operate on a long-term return basis on their investment and credit, due to the environmental reasons, and there is always a risk of non-payment and the reduction of value in credit extension and investment. So, it is vital for the banking sector to adhere to certain safe procedures for the environmental assessment of the projects before providing them funds. There are some researches in the past that have proven a positive correlation between financial performance and environmental performance (Hamilton, 1995). Thus, it is more important for the banks and other financial institutions in the context of environmental performance whether to invest in companies or advise the client to do so. The environmental management has to follow different rule formation for conservation of the resources like the Clean Water Act, Toxic Substance Control Act, and Clean Air Act. All these are recent environmental liabilities for a banking institution. Adopting all these principles will be beneficial for the banking sector and to the financial institution as well as to consumers and also their stakeholders.

8.2   STUDY RATIONALE

Earlier also the environmental sciences existed and society had environmental concerns at large. They took into consideration whether they were degrading the environment or not. The role of environmental sciences changed with the realization that humans induced stresses on various environments and indulged in taking a dangerous toll: Some types of human activities were understood as not being sustainable.

However, in some situations – and with some research questions – it can be harder to see the links between the two. These gaps become particularly apparent in reflecting on the role played by humans and their impacts on the environment. Knowing how current human activities dynamically affect tomorrow's environment is a key focus of study. In this perspective, sustainable development is the end goal, and environmental protection is a subsidiary goal. Many environmental experts focus their work on understanding environmental issues with little or no research focus on (human) development of any kind. Hence, *environmental practices* must be included as one of the top agendas in modern-day society.

Green banking entails banks encouraging environmentally friendly investments and giving lending priority to those industries that have already turned green or are trying to go green, thereby helping to restore the natural environment. The United States suggests if every household were able to switch to paperless bank billing, this would save an estimated 16,500,000 trees per year and gain almost 2,145,000 tonnes of oxygen per year. *International Institute Sustainable Development Bulletin (IISD)* explains that by taking these measures, the health impact, like damage to nervous, immune, and reproductive systems and abnormalities in infant and child development, can be prevented.

Environmental viability is not very independent from commercial viability. That is why some banks have begun considering environment viability along with commercial viability while deciding on lending decisions. The reason being that if a debtor company goes out of business due to environmental issues, the bank is exposed to credit, legal, and reputational risks that hinder the commercial viability of the project as well. However, unlike the foreign banks, Indian banks have a long way to go in this area. Banks have started using renewable energy, which also contributes to employment as IRENA Renewable Energy and jobs. Annual Review 2015 reported that the green energy industry employed 7.7 million people in 2014, up 18 percent from 2013.

The primary objective of study is *to study* the concept of sustainability through green banking and its impact on the environment

8.3   METHODOLOGY

This study is based on secondary data. For the fulfilment of the above-stated objective, a literature review has been done about the impact of green banking on customers, employees, and the organization. This study helps in understanding the relationship of how organizational things are connected with society.

8.4   INDIA'S ROADMAP FOR SDG-13 IN COLLABORATION WITH THE GREEN BANKING MISSION

According to a survey by Futurescape, 60 percent of India's topmost hundred companies comprise Sustainable Development Goals in their long-term strategic plans, and primarily the top ten organizations align their business goals with International Goals. Green banking mission by the Government of India is a major component of India's overall mission for banking activities. The Government of

India launched various schemes under the green banking mission, which made the lives of the citizens a lot easier when it came to keeping track of the things they need to be aware of. Initiatives include going online, card-based transactions, green finance, green infrastructure, and use of power-saving equipment.

Green banking initiatives by Indian banks:

A. Public sector banks.
   1. State Bank of India.
   2. Bank of Baroda.
   3. Punjab National Bank.
B. Private sector banks.
   1. ICICI Bank.
   2. HDFC Bank.
   3. Axis Bank.

Green banking optimizes costs, reduces the risk, enhances the bank's reputation, and contributes to the common good of environmental sustainability. So, it serves both the commercial objective of the bank as well as its social responsibility. Indian Banks can adopt green banking as a business model for sustainable banking. SDG-13 has facilities to strengthen resilience and adaptive capacity to climate-related hazards and natural disasters in all countries; integrate climate change measures into national policies, strategies, and planning; improve education, awareness-raising, and human and institutional capacity on climate change mitigation, adaptation, impact reduction, and early warning; implement the commitment undertaken by developed-country parties to the United Nations Framework Convention on Climate Change to a goal of mobilizing jointly $100 billion annually by 2020 from all sources to address the needs of developing countries in the context of meaningful mitigation actions and transparency on implementation and fully operationalize the Green Climate Fund through its capitalization as soon as possible, and promote mechanisms for raising the capacity for effective climate change-related planning and management in least developed countries and small island developing states, including focusing on women, youth, and local and marginalized communities. The key feature of SDG-13 is improving education, awareness-raising, and increasing human and institutional capacity for climate change mitigation, adaptation, impact reduction, and early warning.

8.5 LITERATURE REVIEW

In this present scenario, environmental management should be the first aim of every organization. Banks have great control of any economy. They can lead businesses, production, as well as the day-to-day life of people. That is why it is becoming more important for banks to adopt green practices. For achieving sustainability, banks should not only make their products green but also adopt environmentally friendly practices in their operations (Ginovsky, 2009). Green banking is an umbrella that covers economic, environment, and social dimensions. Green practices encourage innovation and improve resource utilization (Menon & Shivdas, 2017); therefore, banks are introducing a paperless and technology-oriented process for most of their operations while not losing present customers (Perryman et al., 2016). According to RBI (2014), green banks are those who are using their physical and IT infrastructure to the fullest by having a minimum impact on the environment. Banks are also getting ratings by RBI on the bases of their carbon footprints, recycling, and reuse of materials. Banks should not only make a budget for green finance, green projects, green marketing, and capacity building but also ensure their application. Banks should prepare a green banking and sustainable report, which includes past performance, present development, and future plans. With the underlying notion of sustainability being the ability to meet the needs of today while protecting and enhancing opportunities for the future, the industry needs to focus further on achieving more sustainable practices (Dickson, 2010). Consumption of natural resources adds pressure on the environment because banks have an association with the majority of the population of a country. Banks have a specific contribution to the sustainability development of society as they hold an intermediation position between depositors and borrowers. The financial sector invests capital in other sectors, which will impact the environment by producing waste as well as society and the economy by inefficient use of resources. Sudhalakshmi and Chinnadorai (2014) present the status of Indian Banks with respect to green banking and state that though the going green mantra is essential for emerging economies like India, significant efforts have not been taken. Banks are required to include their green aspect in the lending principle. Every step taken today will mean a better global environment in the future. So, a policy measure to promote green banking is needed in India.

Indian banks are running behind time in adoption of this green phenomenon. Serious steps are required to be taken in this regard.

According to RBI, green banking will make the internal bank processes, physical infrastructure, and IT infrastructure as effective and efficient as possible, with zero or minimal impact on the environment. They introduced green rating standards for Indian banks, which are termed Green Coin Ratings. Under this rating system, banks are judged on the basis of carbon emissions from their operations and on the amount of recycling, refurbishment, and reuse material being used in their building furnishings and in the systems used by them like servers, computers, printers, networks, etc. They are also being judged on the amount of green projects financed by them, and rewards or recognitions are given to borrowers for turning their businesses greener.

Furthermore, climate-related risks are causing physical and transitional risks for the financial sector. To mitigate the negative impacts, central banks, supervisors, and policymakers started undertaking various green banking initiatives, although the approach taken so far is slightly different between developed and developing countries (Park & Kim, 2020). Banks' practices have a direct impact on the environment through effective use of resources and energy (Ercumen et al., 2017). Going green means a bank needs to adopt sustainable practices in all parts of life. Green banking focuses on all stakeholders as it works with government, NSOs, the central bank, customers, and business communities (Bennon & Sharma, 2018). Though the banking industry has never come up in the polluting industry, at present increasing customers and physical facilities like air conditioner is increasing the pollution level in banking. Now, this is the time banks need to take proactive actions regarding their sustainable development (Bhardwaj & Malhotra, 2013). Banks should focus on both internal and external stakeholders. Banks should give preference to green projects before granting loans and other funds.

The study also showed that a barrier to green growth could be the slower rate of technology advancement, financial innovative products, and widespread lack of social and ecological conscience among banking firms. Volz (2018) found sustainable banking where investment and lending decisions are made based on environmental monitoring and risk assessment to fulfil sustainability criteria along with insurance services that address environmental and climatic risk which are significant components of green finance. Bukhari, Hashim, and Amran (2020), while pivoting on a green banking adoptions model based on environmental and social governance, considered where affinity of variables impacted environmental sustainability. This study found that the process is influenced by a variety of environmental factors, and banks can wangle the adoption by applying certain operations in a consecutive and analogous manner. Alsayegh, Abdul Rahman, and Homayoun (2020) claimed that the idea of sustainable banking entails using green banking techniques to take ethical, social, and environmental concerns into account. Khairunnessa, Vazquez-Brust, and Yakovleva (2021) described that the Bangladeshi banks, through their investments in numerous environmentally friendly projects, lessen the negative consequences of climate change and play a vital part in the nation's economic sustainability. Additionally, banking institutions play a significant role in financing numerous industrial projects that could have significant detrimental social or environmental effects. Zheng, Siddik, Masukujjaman, outlined that green financing is seen as a crucial component of sustainable banking, having a significant influence on the growth of an eco-friendly economy and the industry generally. Therefore, it can be said that in enhancing the sustainability practices of the financial sector, the banking sector should focus on ensuring funding for environment-conscious projects through financially viable banking in order to enhance the competitive edge of banks, generate more earnings, improve existing assets, and save on invested capital and other costs. Until recently, green banking appeared to be merely an idea, and environmental concerns did not appear to be particularly relevant to a bank's operations. Initially, a bank evaluating a client's environmental suitability would have been regarded as intruding into their private affairs. As a result, there is a dearth of literature pertaining to green banking in India (Sharma & Choubey, 2022). Not much research has been conducted on the role of green banking in SD overall, green banking practices by SBI (India), Maybank (Malaysia), and the contribution to achieving UN SDGs for the country. Maybank of Malaysia has been first in line for the study due to its top commercial activities in Malaysia, and the State Bank of India is preferred for this review owing to being the first bank to focus on green banking initiatives. Kaur and Sandhu (2019) said most of the studies conducted on green banking predominantly concentrated on green banking practices or on perception of customers or bankers. This gap justifies the need to investigate the problem stated. Therefore, the present study is an endeavour and an attempt to fill the research gap in this regard.

8.5.1 GREEN BANKING PRODUCTS AND SERVICES

1. **Green Deposits:** Banks can offer higher rates on CDs, money market accounts, checking accounts, and savings account if customers opt to conduct their banking activities online.

2. **Green Mortgages and Loans:** A bank can offer green mortgage with better rates or terms for buyers of energy-efficient houses. Some green mortgages allow home buyers to add as much as an additional 15 percent of the price of their house into loans for upgrades including energy-efficient windows, solar panels, geothermal heating, or water heaters. The savings in monthly energy bills can offset the higher monthly mortgage payments and save money in the long run. An Energy Efficient Mortgage (EEM) is a type of HUD-approved green mortgage that will credit you for your home's energy efficiency in the mortgage itself. Many home improvements also qualify for the energy tax credit. Anyone undertaking an energy-saving house project should shop around for a bank that offers a special rate for a green mortgage or loan.

3. **Green Credit Cards:** A green credit card allows cardholders to earn rewards or points that can be redeemed for contributions to eco-friendly charitable organizations. These cards offer an excellent incentive for consumers to use their green card for their expensive purchases. Imagine the millions of dollars that could be raised for worthwhile environmental groups if green credit cards really took off.

4. **Green Rewards Checking Accounts:** Rewards checking accounts may pay a bonus rate for customers who go green. Customers can earn higher checking account rates if they meet monthly requirements that might include receiving electronic statements, paying bills online, or using a debit or check card. With this type of banking product, higher rates and eco-friendly living go hand-in-hand.

5. **Mobile Banking:** By using it, customers can check balances, transfer funds, or pay bills from the phone. It also helps to save time and energy of the customers.

6. **Online Banking:** It means customers can perform most of their banking-related functions without visiting the banks personally. For this, the customer must possess an Internet banking ID and a password provided by the bank in which the customer has an account. Online banking includes the use of credit cards, debit cards, online bill payment, and electronic fund transfer.

7. **Green CDs:** With a Green Bank CD, customers can earn a guaranteed rate of interest over a term that the customer can choose. Banks offer a wide variety of certificates, ranging from seven days to five years. Interest on these accounts can be compounded quarterly, paid monthly by check, or transferred to a green banking deposit account.

8.5.1.1 Green Banking and the Customer

Green banking is beneficial for the customer because the customer can save time, fuel, energy, paper, water, and money. Environmental practices like Internet banking, e-statements, online bill payments, etc. (Rahman & Perves, 2016) help the customer to make a contribution towards the environment by reducing their carbon emission while performing banking tasks (Islam & Das, 2013; Sudhalakshmi & Chinnadorai, 2014; Deka, 2015; Rahman & Perves, 2016). Green banking helps by improving the operational process, increasing the use of technology, and making customers familiar with this (Biwas, 2011). Even the people who are using technology for their day-to-day banking practices are not aware of green practices. Many times, when customers were asked about green practice, they just knew about online transactions. They were not aware of other practices like solar energy, bonds on environmental protection, and government relaxation for green practices. In a traditional way, banks were using a lot of paper in cash deposit, payment in the form of cheques, and other forms. Now, banks are using various technology like mobile banking, Internet banking, etc., for their services. These services make the banking process fast and flexible with the reduced use of paper and cost reduction (Vijayasarathi & Velmurugan, 2016). When the customer goes for banking services, after looking for a safe and secure place, they are also a bank found more responsible towards society and are involved in contributing toward society development Amirul (Muhamat & Nizam bin Jaafar, 2010).

Various studies found that green practices in any organization lead towards customer satisfaction and loyalty. It was also found that organizations incorporated green practices due to market pressure, to gain market share and competitive advantages, to improve brand image, and to add more value to their products. Green banks are the banks that are more reliable in terms of safety, providing value-based services, and being environmentally responsible (Muhamat et al., 2010). Various studies found that green banking practices help reduce risks for the customer and banks (Dharwal & Agarwal, 2013). Banks are adopting green practices for reducing risk by the use of technology and gaining more profit for investors. With such aims, banks try to implement more and more green practices, which would make services easier through the use of technology to attract more customers and investors (United State Department of Energy Report, 2017). Green banking contributes in various terms to the economy like spreading renewable energy to consumers and creating employment, and also there are various tax relaxations that are given to businesses who are operating green practices. When the organization uses the environment with their marketing practices, this with help in attracting customers and creating customer loyalty. Green banking not only helps in sustainability of the economy and environment, but also improves the image of banks in the eyes of customers. Post-purchase behaviour of the customer is affected by the image the customer perceives about the organization and product, which will lead to customer loyalty. A Study on Effective Utilisation of Green Channel Counter with Special Reference to City Union Bank Pvt. Ltd, Vellore Branch.

In banking, the most important operations, like cash deposit, cash withdrawal, and inter banking transfer, can be done by the way of writing channels and using cheque leaves. The banks have implemented certain technology services to cut down the usage of more paperwork and reduce transaction costs. Green Channel Counter avoids as much paperwork in their banking transactions as possible. It adopts and implements environmental standards for lending, which is really a proactive idea that would enable eco-friendly business practices that would benefit future generations. The customers felt that the cash handling charges were collected and even remit cash to the Xpress Desk. The Green Channel Counters help the customers to have convenient and comfortable banking transactions. Sustainable Development Goal 13 (SDG-13, or Global Goal 13) is about climate action and is one of 17 Sustainable Development Goals established by the United Nations General Assembly in 2015. The official mission statement of this goal is to "Take urgent action to combat climate change and its impacts."[1] SDG-13 has five targets, which are to be achieved by 2030. They cover a wide range of issues surrounding climate action. The first three targets are "output targets": strengthen resilience and adaptive capacity to climate-related disasters; integrate climate change measures into policies and planning; build knowledge and capacity to meet climate change. The remaining two targets are "means of achieving" targets: to implement the UN Framework Convention on Climate Change and to promote mechanisms to raise capacity for planning and management. Along with each target, there are "indicators" that provide a method to review the overall progress of each target, along with SDG-13 as a whole. The United Nations Framework Convention on Climate Change (UNFCCC) is the primary international, intergovernmental forum for negotiating the global response to climate change.

8.5.1.2   Green Banking and Employees

Implementation of green practices depends on the day-to-day practices of employees. The workplace should have environmentally sound policies, and employees should work with full efficiency of resources. This also means that organizations should maintain a good human resource department that can help in developing employees' attitude towards green practices. Green baking is not only important for the environment but also for reducing daily work pressure and reducing stress that leads to better work quality. Green practices related to employees have a significant impact on the bank's environmental performance, and they will help in creating organizational citizenship (Shaumya & Arulrajah). Green initiatives in the organization would make everyday life of employees easy and less loaded and also provide greater efficiency, lower costs and help in making good relation among employees that result in for of better organization performance (Mathapati, 2013). HRM is the most effecting function for an organization. The HR manager as a stakeholder of the organization understands their duties towards the environment, but it is not an easy task to incorporate green practices in every employee (Fenwick & Bierema, 2008).

Due to these conflicting situations, green HRM has come into existence for taking proactive actions towards the environment and saving us from future disaster (Bhatti et al., 2012). Increasing environmental issues like climate change, greenhouse effect, etc., bring everyone's attention towards green practices, and for adopting green practices it is necessary that each and every employee of the organization should be involved in environmental practices not only at the workplace but in their personal life also. Many studies have found that organizations' productivity and financial performance are directly affected by their HR practices.

Banks can go green by just adopting a few environmentally friendly practices in day-to-day life through going online and reducing paperwork. These practices also include doing e-recruitment, e-training and development, creating employee awareness towards environmental practices, and also doing green practices like planting trees as a social responsibility (Miah et al., 2020). These practices result in improving the effectiveness, efficiency of organization, and also improve performance of employees, improve work-life balance, reduce the cost and wastage of operating, and also reduce the turnover ratio of the organization. Green banking also helps to reduce carbon emission of every employee as well as comfort in the job by practicing electronic filing, job-sharing, teleconferencing, audio-video conferencing, work flexibility, etc.

8.5.1.3   Green Banking and Organization

Green banking involves the process of implementing environmentally friendly practices like online channels, providing green loans, and other CSR activities, etc. *Green financial product and carbon reduction policies include paperless office, energy saving, mass transportation and no transportation practices, green building, and CSR activates (Sarita, 2012).* Green loans, green projects, and environmental training posed substantial impact on Bank's environmental performance. Banks and government initiatives for green practices are important for banks reputation and customer awareness. Organizations have started conducting energy audits, surveys, recycling the resources, reduce transportation, save water, and other environmental practices that will result in cost reduction, increasing market shares, and also satisfied all stakeholders. To safeguard and accelerate the growth rate of the economy, banks need to adopt a proactive nature towards the environment. Green practices help in attracting customer, build market image and get greater legitimacy. Making a green Initiative not only is good for nature but also beneficial for the organization as it helps to earn carbon credits, decrease costs, and give a competitive advantage (Yadav & Swaroop Pathak, 2013). It also helps in gaining efficiency in operation, lowering vulnerability, and reducing costs. Go Green is a wide concept that would lead the growth, variation, and innovation in an organization (Sudhalakshmi & Chinnadorai, 2014). Green banking is beneficial for both customers and banks because customers can save their time, fuel, energy, paper, water, and money. Similarly, banks have to bear less cost and reduce workload (Deka, 2015). Green practices also build a positive image in the market (Shaumya & Arulrajah, 2018). Organization performance is directly affected while practicing green banking (Bhardwaj & Malhotra, 2013). That is a great motivation for banks; they should go for green banking and save our ecosystem (Sahoo & Nayak, 2007). Green practices in banks will not only improve environmental conditions but also improve the operations: reducing error, reducing deception, and also reducing the cost of operations. Green practices are not a cost for environmental protection but in the long run, green practices will be beneficial for businesses, banks, and the economy (Biwas, 2011). As a part of society, organizations have some responsibility towards society and the environment. When organizations make their decision with consideration of the environment and society, they are not only doing their responsibilities but also making their operations smooth and effective. Through the use of mobile, Internet, and other technology banks can improve the quality of assets and can reduce their credit risk and liabilities risk. Although there are various researches that are focusing on the customer perspective towards green practices, very few researchers have focused on the banker's perspective (Aizawa & Yang, 2010; Fernando & Fernando, 2016).

Environmentally friendly practices related to employees have a significant impact on organizational environmental performance and will help in creating organizational citizenship (Shaumya & Arulrajah). Environmental and sustainable initiatives in the organization would make the everyday life of employees easy and less loaded and also provide greater efficiency, lower costs, and help in making good relations among employees that result in better organization performance (Mathapati, 2013).

8.5.1.4   Organizations Towards Self-Reliance through Environmentally Sustainable Methods

Implementing environmentally friendly practices like online channels, providing green loans, and other CSR activities, etc., are some of the activities organizations can adopt at their level to meet the goals. Organizations have started conducting energy audits, conducting surveys, recycling resources, reducing transportation, saving water, and other environmental practices that will result in cost reduction, increasing market shares, and also satisfied stakeholders (Ayeswarya, 2017). To safeguard and accelerate the growth rate of the economy, organizations need to adopt a proactive nature towards the environment. Sustainable practices help in attracting customers, building market image, and getting greater legitimacy (Zhixia, Husan, Mujafary, & Begum, 2018). Taking Environmental Initiative is not only good for nature but also beneficial for the organization as it helps in earning carbon credits, decreases costs, and gives a competitive advantage (Yadav & Swaroop Pathak, 2013). It also helps in gaining efficiency in operation, lowers vulnerability, and leads to cost reduction. Go Green is a wide concept that would lead the growth, variation, and innovation in an organization (Sudhalakshmi & Chinnadorai, 2014). Sustainable practices are not cost for environmental protection, but in the long run, these practices will be beneficial for agriculture, the economy, business, and industries (Biwas, 2011). As a part of society, organizations have some responsibility towards society and environment. When organizations make their decision with the consideration of the environment, sustainability, self-reliance, and society, they are not only doing their responsibilities but also making it's operations smooth and effective (Bihari & Chandra, 2010). The full text of Target 13.a is: "Implement the commitment undertaken by developed-country parties to the United Nations Framework Convention on Climate Change to a goal of mobilizing jointly $100 billion annually by 2020 from all sources to address the needs of developing countries in the context of meaningful mitigation actions and transparency on implementation and fully operationalize the Green Climate Fund through its capitalization as soon as possible." This target only has one indicator: Indicator 13.a is the "Amounts provided and mobilized in United States dollars per year in relation to the continued existing collective mobilization goal of the $100 billion commitment through to 2025." Previously, the indicator was worded as "Mobilized amount of United States dollars per year between 2020 and 2025 accountable towards the $100 billion commitment."

This indicator measures the current pledged commitments from countries to the Green Climate Fund (GCF), the amounts provided and mobilized in United States dollars (USD) per year in relation to the continued existing collective mobilization goal of the US$100 billion commitment to 2025.

Currently, according to the United Nations, data for this indicator are not yet developed. Regarding funding, by December 2019, 81 countries submitted 83 proposals totalling $203.8 million, requesting support from the GCF. There was an increase of $681 billion from 2015 to 2016 with regard to global climate finance.[11] Renewable energy received high levels of new private investment. This represents the largest segment of the global total.[11] These financial flows are relatively small in relation to the scale of annual investment needed for a low-carbon, climate-resilient transition. In April 2018, 175 countries ratified the Paris Agreement, and 168 parties had communicated their first nationally determined contributions to the UN framework convention on Climate Change Secretariat.[42] As of March 2020, 189 countries had ratified the Paris Agreement, and 186 of them – including the European Union – have communicated their Nationally Determined Contributions (NDC) to the Secretariat of the United Nations Framework Convention on Climate Change.

8.6   DISCUSSION

This study takes a more comprehensive assessment of sustainability survey research, although there have been other articles published with a similar objective. In this study, the attempt has been done to discuss various literature about green practices in the banking industry and its impact on customers, employees, and the organization. Every organization is a part of society, and they have some responsibility towards society. The world is burning in fires of pollution. Now an organization's first duty is creating sustainability in the environment, society, and businesses also. Adopting environmentally friendly practices not only makes our environment sustainable but also contributes to the goodwill of banks. Green practices help banks in reducing the cost of operation, help in attracting more customers and investors, and also create loyalty in the eyes of customers. These technology-oriented practices also help in customer services like online banking and mobile banking. Due to the ease of the process, customers feel more satisfied as well as safe and secure. These processes are reducing customers' costs and processing time. Customers also feel proud to be associated with the banks, which is having a good image in the market.

Green banking is not only beneficial for the organization and customer. Even the banks' employees feel proud while working in a reputed organization. These green practices also reduce the workload of employees and help reduce stress of record maintenance and occurrence of errors in work as automation makes it more secure and reliable and reduces the risk of credibility. From the literature and secondary data from published sources, it is found that saline soils are going to increase by 50 percent up to 2050, and land degradation has affected almost every country of the world.

The Intergovernmental Panel on Climate Change's (IPCC's) fall 2018 report finds that to meet the goal of limiting global warming to no more than 1.5 degrees Celsius, investments in low-carbon energy technology and energy efficiency will need to increase by roughly a factor of five by 2050 compared with 2015 levels. This will require an unprecedented mobilization and redirection of domestic and international capital. In short order, successful strategies will have to be adapted and scaled, and new financial instruments will have to be deployed to apportion risk in novel ways.

8.7 CONCLUSION

Environment degradation is increasing day by day; so are concerns towards the environment increasing in everyone's mind. Environmental degradation is a big concern for not only our generation but also for upcoming generations. Climate change is expected to accelerate and is no longer considered only an environmental threat because it affects all economic sectors. This concern has become the first priority of every country. Banks are having a large participation in any economy as they can affect everything from the common man to large businesses. The banking industry has a direct influence on every single individual, whether they are from inside of the organization or outside the organization. Policymakers are realizing the importance of adopting green banking practices as an intervention measure for environmental sustainability. The efforts can only be successfully implemented if the code of conduct for the same is practiced as a regular approach. This study attempts to analyze green practices in banks and also attempts to identify motivating factors for the customer, employee, and organization to practice these green practices. Many banks have themselves started practicing green standards, and this is one prime step to strategize environmental and social reforms. These motivating factors can contribute to improving the lifestyle of every individual, economy, society, and environment with standards that need to be developed to measure green banks' performance for a major change in climate.

REFERENCES

Aizawa, M., & Yang, C. (2010). Green credit, green stimulus, green revolution? China's mobilization of banks for environmental cleanup. *The Journal of Environment & Development, 19*(2), 119–144.

Alexander, K. (2016) Greening banking policy. In: *Support of the G20 Green Finance Study Group.*

Alsayegh, M. F., Abdul Rahman, R., & Homayoun, S. (2020). Corporate economic, environmental, and social sustainability performance transformation through ESG disclosure. *Sustainability, 12*, 3910.

Bahl, S. (2012). Role of green banking in sustainable growth. *International Journal of Marketing, Financial Services and Management Research, 1*(2).

Bennon, M. , & Sharma, R. (2018). *State of the practice: sustainability standards for infrastructure investors.* Available at SSRN 3292469.

Bhatti, N., Maitlo, G. M., Shaikh, N., Hashmi, M. A., & Shaikh, F. M. (2012). The impact of autocratic and democratic leadership style on job satisfaction. *International Business Research, 5*, 192.

Bhardwaj, B. R., & Malhotra, A. (2013). Green banking strategies: Sustainability through corporate entrepreneurship. *Greener Journal of Business and Management Studies, 3*(4), 180–193.

Bihari, S. (2011) Green banking-towards socially responsible banking in India. *International Journal of Business Insights and Transformation*, *4*(1) October 2010 – March 2011.

Biswas, N. (2011). Sustainable green banking approach: The need of the hour. *Business Spectrum*, *1*(1), 32–38.

Bukhari, S. A. A., Hashim, F., & Amran, A. (2020). Green Banking: A road map for adoption. *International Journal of Ethics and Systems*, *36*, 371–385.

Chowdhury, S. H., Roy, S. K., Arafin, M., & Siddiquee, S. Green HR practices and its impact on employee work satisfaction-A case study on IBBL, Bangladesh.

Crane, A., Matten, D., & Moon, J. (2010). The emergence of corporate citizenship: historical development and alternative perspectives, *Corporate Citizenship in Deutschland: Gesellschaftliches Engagement von Unternehmen. Bilanz und Perspektiven* (pp. 64–91).

Dharwal, M., & Agarwal, A. (2013). Green banking: An innovative initiative for sustainable development. *ACCMAN Institute of Management Article*, *2*(3), 1–7.

Dickson, C. (2010). Promoting sustainable event practice: The role of professional associations. *International Journal of Hospitality Management*, *29*(2), 236–244.

Ercumen, Ayse, Pickering, A. J., Kwong, L. H., Arnold, B. F., Parvez, S. M., Alam, M., Sen, D., Islam, S., Kullmann, C., Chase, C., Ahmed, R., Unicomb, L., Luby, S. P., & Colford, J. M. (2017). Animal feces contribute to domestic fecal contamination: Evidence from *E. coli* measured in water, hands, food, flies, and soil in Bangladesh. *Environmental Science & Technology*, *51*, 8725–8734.

Fenwick, T., & Bierema, L. (2008). Corporate social responsibility: issues for human resource development professionals. *International Journal of Training and Development*, *12*, 24–35.

Ginovsky, J. (2009). Green banking - Inside and out. *Community Banker*, 30–32.

Gupta, J. (2015). Role of green banking in environment sustainability-A study of selected commercial banks in Himachal Pradesh. *International Journal of Multidisciplinary Research and Development*, *2*(8), 349–353.

Hamilton, J. T. (1995). Pollution as news: Media and stock market reactions to the toxics release inventory data. *Journal of Environmental Economics and Management*, *28*, 98–113.

https://www.thrall.org/special/goinggreen.html

https://www.newbusinessage.com/Articles/view/2804

https://www.thehindu.com/society/what-does-self-reliance-really-mean-amazing-stories-emerge-from-indias-villages/article31756580.ece

www.Ecomicstimes.com

https://www.mapsofindia.com/my-india/business/top-10-largest-public-sector-banks-in-india- 2019#

https://www.oliveboard.in/blog/private-banks-india/

http://www.cwejournal.org/wp-content/uploads/2015/09/Vol10_No3_Gree_Kan_Fig2.jpg

IDRBT (2013). Greening banking for Indian banking sector.

Ibe-Enwo, G., Igbudu, N., Garanti, Z., & Popoola, T. (2019). Assessing the relevance of green banking practice on bank loyalty: The mediating effect of green image and bank trust. *Sustainability*, *11*(17), 4651.

Islam, M. S., & Das, P. C. (2013). Green banking practices in Bangladesh. *IOSR Journal of Business and Management*, *8*(3), 39–44.

Kaur, K. , & Sandhu, V. (2019). Green initiatives in banking sector: A study of the State Bank of India (SBI). *ZENITH International Journal of Multidisciplinary Research*, *9*, 111–120.

Kapoor, N., Jaitly, M., & Gupta, R. (2016). Green banking: A step towards sustainable development. *International Journal of Research in Management, Economics and Commerce*, *7*, 69–72.

Khairunnessa, F., Vazquez-Brust, D. A., & Yakovleva, N. (2021). A review of the recent developments of green banking in Bangladesh. *Sustainability*, *13*, 1904.

Mathapati C. M. (2013). Green HRM: A strategic facet. *Tactful Management Research Journal*, *2*, 1–16.

Mebratu, D. (1998). Sustainability and sustainable development: Historical and conceptual review Environ. *Impact Assessment Review*, *18* (1998), 493–520.

Miah, M. D., Rahman, S. M., & Mamoon, M. (2020). Green banking: The case of commercial banking sector in Oman. *Environment, Development and Sustainability*, 1–17.

Morelli, J. (2011). "Environmental sustainability: A definition for environmental professionals," *Journal of Environmental Sustainability*, *1*(1), Article 2. DOI: 10.14448/jes.01.0002 http://scholarworks.rit.edu/jes/vol1/iss1/2

Muhamat, A. A., & Nizam bin Jaafar, M. (2010). The development of ethical banking concept amongst the Malaysian Islamic banks. *Norfaridah, The Development of Ethical Banking Concept Amongst the Malaysian Islamic Banks (February 25, 2010). iCAST*, 24–25.

Nath, V., Nayak, N., & Goel, A. (2014). Green banking practices – A review. *International Journal of Research In Business Management*, *2*(4), 45–72.

Park, H., & Kim, J. D. (2020). Transition towards green banking: Role of financial regulators and financial institutions. *AJSSR*, *5*, 5. 10.1186/s41180-020-00034-3

Perryman, A. A., Fernando, G. D., & Tripathy, A. (2016). Do gender differences persist? An examination of gender diversity on firm performance, risk, and executive compensation. *Journal of Business Research*, *69*, 579–586.

Rahman, F., & Perves, M. M. (2016). Green banking activities in Bangladesh: An analysis and summary of initiatives of Bangladesh Bank. *Research Journal of Finance and Accounting*, *7*(10), 6–7.

Sahoo, P., & Nayak, B. P. (2007). Green banking in India. *Indian Economic Journal*, *55*, 82–98.

Sharma, M., & Choubey, A. (2022). Green banking initiatives: A qualitative study on Indian banking sector. *Environment, Development and Sustainability*, *24*, 293–319.

Shaumya, K., & Arulrajah, A. A. (2018). The impact of Electronic Human Resource Management (e-HRM) practices on bank's environmental performance. *Journal of Business Studies*, *2*, 33–50.

Sudhalakshmi, K., & Chinnadorai, K. (2014). Green banking practices in Indian banks. *International Journal of Management and Commerce Innovations*, *2*(1), 232–235.

Veena, K., & Nayana, N. (2017). A study on customer perception towards green banking – A special reference to State Bank of India - Bangalore and Mysore City. *Journal of Exclusive Management Science*, *6*, 02.

Vijayasarathi, V., & Velmurugan, G. (2016). A study on usage of self service Kiosk with special reference to City Union Bank Pvt Ltd, Vellore Branch. *Asian Journal of Research in Banking and Finance*, *6*(3), 26–30.

Volz, U. (2018). *Fostering green finance for sustainable development in Asia*. Tokyo: Asian Development Bank Institute.

Weber, O. (2014). Social banking: Concept, definitions and practice. *Global Social Policy*, *14*(2), 265–267.

Yadav, R., & Pathak, G. S. (2013). Green marketing: Initiatives in the Indian context. *Indian Journal of Marketing*, *43*(10), 25–32.

9 Managing Employees in Private Organizations
Dilemmas and Strategies for Managers

Himani Grewal and Smrita Jain
Department of Management, Moradabad Institute of
Technology, Moradabad

9.1 INTRODUCTION

Imagine a case where your child has been kidnapped and you are chasing the kidnapper. You are driving carefully, but somehow you hit a pregnant woman crossing the road. Who are you going to save: your child or the lady alongside the road?

Now assume another case in the context of a workplace. You are working as a manager in a renowned company and leading a team of ten members. Among those ten members, there is one female and nine males. The female member of your team is regularly bullied by the male members. Even though you have noticed the behaviour of other members towards her, you have not taken any action as the female member didn't complain to you on this matter. You do not want to highlight this matter because you are about to reach your monthly target. What will you do?

Imagine another case at work. Your firm is not doing well due to COVID, and you are a manager leading a team of ten well-qualified and experienced professionals who have been working with the company for a decade. You are advised to condense your team into five members only. Your team has worked exceptionally to reach the goals and contributed to the firm's development. Now every single team member has his or her own reasonable personal challenge, like one is divorced with two children, one has lost his father, another one's parent is ill and hospitalized, and several have their home affected by the same kind of genuine problems. But the management does not have enough funds to maintain a big office now. So, among them, who are going to keep with you?

These kinds of situations are very normal in our daily life. All people deal with the process of decision making in their life when a number of choices are available to them and the best alternative should be selected. Our ordinary behaviour patterns and our professional behaviour frequently diverge. In situations where we might lose our temper outside of work, we must be amiable and polite instead. We generally keep our feelings (and occasionally our ideas) to ourselves at work, and

DOI: 10.1201/9781003365525-9

approach situations with more professionalism at the workplace. These types of conditions generate a number of dilemmas. When it comes to a workplace, the managers are responsible for handling these situations accordingly. According to Van Gramberg and Teicher (2006), managers are important to resolve the conflicts between employees. So, the managerial role is crucial in any organization to deal with dilemmas.

A dilemma is generally defined as a study of a certain kind of predicament, which is experienced by each and every person at some point in time in his or her life. This happens when making a decision is difficult, and occasionally impossible, when the important options we have or the important commitments we must fulfill seem to be so evenly balanced. It originally referred to two and only two options, each having undesirable outcomes, and having a clear meaning.

Wallace (2015) defined the term *dilemma* as "a conflicting state." Cambridge Dictionary defines *dilemma* as "a situation in which a difficult choice has to be made between two different things you could do." This word is a combination of the Greek words "di" and "lemma." The former denotes "two," and the latter, "proposition." This phrase is described as "a scenario in which a difficult choice must be made between two or more alternatives, especially ones that are equally disagreeable" by the Oxford English Dictionary. Common usage may emphasize unfavourable or negative alternatives, whereas innovation-related decisions are more frequently made between two positive options. Individual against group, small details versus the big picture, control versus cooperation, and constancy versus change are a few examples of common conundrums.

People deal with the decision-making process every day, whether it is in their personal or professional lives. They typically have a few options to choose from. Individuals must therefore make a choice as a result. The objective is to choose the best candidate, yet what constitutes the finest varies for numerous individuals, organizations, and countries. Additionally, it can be challenging for the decision maker to decide which option to choose because the possibilities usually conflict. Every enigma, in essence, contains an ethical conundrum (ED), i.e., whether the choice is ethically good or awful, fair or unfair.

Cuban (1992) distinguished between two types of conflicts based on how soon they can be handled: "dilemma" and "problem." In contrast to a dilemma, which arises when one must choose between two "rights" or "wrongs," a problem is "any situation in which one must make a decision between right and wrong." In other words, by making the proper decisions, situations of any complexity and ugliness can be resolved. Dilemmas, on the other hand, are unsolvable since they call for a decision between two or more desired but conflicting values. Since problems that may be solved simply are not dilemmas, they are also defined as "cognitive knots."

Educational experts have defined the word "dilemma" in a variety of ways. Instead of listing definitions, it would be better to concentrate on the concepts that underlie these definitions in this instance. Nearly all definitions identify a dilemma as having one or more of the following three traits: The dilemma presents a decision between two values that are, or appear to be, mutually exclusive (Wallace, 2015).

This article discusses the essence of dilemmas, different kinds of dilemmas in the workplace, and some rules to resolve dilemmas. With the help of two real-time cases, it has been shown that dilemma is constantly present in experiences that can occur at any time when you are confused to say yes or no to any event or to choose between right or wrong.

9.2 ESSENCE OF DILEMMA

The phenomenon of management dilemma is briefly described in a small number of publications in the discipline of management. Some, like Neuberger, illustrate the duality of leadership between conflicting objectives on the part of management and personnel. These imply areas of friction between a social system's task-oriented and personnel-oriented design (Neuberger, 1995).

According to Fontin, a dilemma is a particular type of logical conclusion. First of all, a constructive dilemma is a challenge of decision making in which a goal can be accomplished in two different ways, but there is no compelling argument for choosing any one particular method over the other. Second, a destructive dilemma makes choosing between two options impossible. Every alternative is a sensible decision. However, their simultaneous implementation is not possible due to the current management scenario and its circumstances. A circumstance where a decision must be made and where there are two reasonable possibilities with equal but opposing supporting evidence is referred to as a dilemma. It is illogical to draw any conclusions or make a rational decision between these options (Fontin, 1997).

9.3 TYPES OF DILEMMAS IN THE WORKPLACE

Workplace dilemmas can be of different types. It totally depends on the situation an employee is going through at his or her workplace. Employees can experience different kinds of dilemmas, but the ethical dilemma is the most observed dilemma in the workplace. For example, do you say yes or no to a particular job, report an inappropriate event, or lie to senior managers? What do you do when you do not agree with the manager's point but can't say no? When you are not keeping a balance between work and home? These dilemmas can influence the work culture and overall efficiency of the employee.

When presented with a decision that has the "horns of a dilemma" (Peters, 2012), managers frequently have to choose between two logical but seemingly incongruent options that each have pros and downsides (Hülsmann & Berry, 2004). Typically, these options are presented in either/or terms, with either option being able to be argued rationally (Peters, 2012). These managerial conundrums frequently have two seemingly incompatible situations, which are intrinsically paradoxical.

Such conflicts include, for example, those involving exploitation and exploration (Andriopoulos & Lewis, 2009), employee autonomy and managerial control (Gilbert & Sutherland, 2013), organizational stability and transformation (Farjoun, 2010), and working with rivals (Chin, Chan & Lam, 2008). According to Johnson (2012) and Peters (2012), dilemmas are usually described as having two extreme solutions that would result in opposing gains. According to Hülsmann and Berry (2004), management conundrums can arise in circumstances where there are conflicting standards for evaluating success inside the organization. According to Smith and Lewis, the choice will result in a trade-off with potentially unforeseen effects because each horn of the dilemma reflects a separate set of costs and rewards.

Dilemmas can fall under two main categories: one is moral dilemma (doing something for the common good, for personal interest, and for the interest of others) (Rossouw & Van Vuuren, 2006), and another is ethical dilemma (emerging from rights, principles, values, and duties).

9.4   IS THERE ANY RULE TO SOLVE A DILEMMA?

There are some points that one can keep in mind and follow accordingly to find the solution to a dilemma. One can consider the consequences of one's actions before deciding on anything. This process is known as ends-based thinking. Another rule says that one should do what is going to help the greatest number. An action is said to be the right action if it is a universally acceptable moral rule. This is known as rule-based thinking. The next rule says that one should empathize with others in such a way that one wants to be empathized by others, known as care-based thinking.

These three are known as decision-making principles and are helpful in resolving ethical dilemmas that emerge at the time of clashes of two values.

The rule of hierarchy and the rule of choice can also be used to solve ethical and moral dilemmas in which priority to each available alternative can be given to the conditions and the top-most solution would be the solution or greatest option. However, choosing between options is typically influenced by feelings and other factors. The two components of emotions are cognitive and empirical. In cognitive emotions, there can be a positive or negative experience that an actor acquires in addressing ethical difficulties, while empirical emotions are the beliefs regarding the process of resolving ethical dilemmas. These choices will also differ on the basis of urgency of the situation. There can be a difference also why to choose the rule of hierarchy and why choose the rule of choice if it is not getting fruitful results. Rules for resolving the dilemma can be straightforward and complex also (Steps to Solve an Ethical Dilemma, 2008).

9.5   SOME REAL-TIME CASES OF DILEMMAS

9.5.1   CASE 1

"I will talk to you later on. It's already 9.15 am. I have to reach my desk; otherwise, I will be in trouble." Kartik told one of his colleague, heading towards his office.

Kartik was late again for work. "This is the third time you are late in a week. If this is the way you will do the work, you can't continue for a longer period," the HR Manager said.

"You know that I have to travel a long way to come to the office. Moreover, due to my leg I can't walk faster. You have to understand and should empathize with my situation. I am only 15 minutes late, and I try all efforts to come on time, but sometimes the situation is not in my hand," Kartik argued.

The HR Manager replied, "I understand, but you have to follow the norms of the organization as you are a part of that. As per the policy of the organization, after being late for 15 minutes three times a week, half leave will be marked, and we have to follow that."

The HR Manager had already warned him several times about coming late. Besides coming late to the office, he was also not active at his job. He had a problem in his leg. Due to that, he was not able to walk properly. At the initial stage, the HR manager had sympathy with him, but now this was affecting the work, and his productivity was going down. The employees were in favor of Kartik, and they asked the HR Manager if he should be given some privileges.

Amit had been working in the organization for eight years and held a dominant position among the employees. The employees always gave consensus for their opinion. John was a bit soft towards Kartik as he considered him a man with a lot of misfortune. He fought with the HR manager for him and said that management has to support him otherwise all the employees will revolt.

"How can the organization differentiate among employees? If special treatment would be given to him, then some of you will come to me and complain for the favor that Kartik is getting," the HR Manager kept his viewpoint. But as the employees were adamant, they further argued that he is the sole earner, and he has a joint family to take care of. He has two younger brothers and old parents. Whenever anyone has any issue, he has to look after that. Moreover, he lives very far away. If the organization will provide him with a place to stay on the organization premises, it would be very convenient for him, and he can come on time as well as his work will also not suffer.

"Office apartments are for senior profile executives, and giving a stay to the junior employee is against the policy of the organization. Taking such a step would not be beneficial for the longer term as other employees can also demand it by giving various other excuses and the apartments are quite expensive to afford as they are being provided at the time of negotiation and as a perk for the senior level of management," the HR Manager responded in a irritating tone.

Amit immediately replied, "We are not demanding apartments that are for senior managers, but two room sets are also there that are constructed for staff members. That can be given to him, and some amount can be deducted from his salary for that as it will not affect him as he is also paying rent for his house."

The HR Manager now is in dilemma with how to cope with this situation as management can hire a better candidate in place of Kartik, but the employees are on his side.

9.5.2 CASE 2

Leena was not able to sleep properly. Ambiguity in her thoughts was making her restless, and she was thinking again and again whether she was making the right decision.

Leena was working as Marketing Head in XYZ Corporation. She had devoted 15 years to the same organization. She was getting a handsome salary and other benefits, along with having supportive team members to lead. Conditions were favourable, and everything was going well, but this phase couldn't last any longer due to a downfall in sales, increasing competition, and customers shifting to other rival companies. Management decided to stop raises. They committed that when the conditions improved, everyone would get a raise.

Five years had passed, and conditions were improving slowly, but there was no sign of increases in salary. One thing that was stopping everyone from leaving the organization was the healthy working environment and good team members. As Leena was the Head, every member of her team followed her instructions. A cordial atmosphere with good working conditions was binding every member of the organization to each other. In spite of no raise for five years, the retention rate was very high. Leena was respected and was getting proper recognition in all the events and programs of the organization.

Leena was very confused and told her husband, "I don't know what to do. I am satisfied and happy in this organization, but financial growth is also very important. I got an offer from a rival company. They are offering me a 40% raise in the current salary with other perquisites and high designation."

Her husband said, "I will support your decision, but I think a 40% raise is a good one. You should think about this offer."

Leena replied, "I have enquired about the company. A lot of work pressure is there. No doubt this offer will give me growth, and this is a good opportunity to grab as the company is more diversified and will be a learning experience for me."

The next morning, Leena knocked on the door, "May I come in?"

The Managing Director Mr. Rao was having his coffee. "Come in, Leena. How are you?"

"Want to have coffee?" he asked.

"No sir," Leena replied.

"Tell me what I can do for you," he asked, having a sip.

Leena said, "Actually, sir, this is my resignation letter. This organization is like my home, and after spending 15 years of my life here, it is very difficult to make this decision, but I have to make it because if today I don't make it, then for the rest of my life I will not be able to do it. I got a very promising offer from another company, and they are giving me a raise of 40% as well as promoting me to a senior level."

Mr. Rao was shocked to hear that as he knew Leena since the time of her joining and he had always appreciated her sincerity, dedication, and commitment level towards the organization.

He replied, "I understand that due to financial constraints management was not able to give raises in salary, but soon everyone will get that as the conditions are getting better. I know whatever offer you are getting is good, and that much raise we cannot give, but you yourself know that this organization is a great place to work. You are already well acquainted with the culture; every employee has a favourable opinion about you. Management treats you as an asset to the organization. You have to think again before making a final decision, as money doesn't matter. We must consider other factors as well."

It was 11 pm, and Leena was puzzled what to do. Making decisions was not as easy as she thought.

The next morning, Mr. Rao had a discussion with other members of the board.

Mr. Rao said, "She is an asset for the organization and having her go is not good for the department health. I think we have to stop her and must provide at least 30%

increase in the salary." Though all the board members were well aware about Leena's capabilities, they were not sure about this much of a raise.

One member of the board argued, "If we give an increase on the grounds that the employee has given a resignation and we are raising her salary so she won't leave, we will set an example in front of other employees that we can also increase their salary if they give a resignation. I think we should not start this trend."

Other members of the board nodded their heads, showing their consent on this point.

Mr. Rao now had a dilemma of what to do as he didn't want a good employee to go, but whatever the board had decided was also important.

9.6 CONCLUSION

Dilemmas are a reality for organizations and should not be avoided by managers. A healthy working atmosphere can be created with the condition that dilemmas should not affect the mental health of the employees. If such a problem is identified, there is a requirement to implement a proper system so that such conditions can be managed. With the help of the above-mentioned cases, it can be easily observed that all sectors are facing such issues, and if these are not resolved at the right time, organizations can suffer. This article has focused on the various aspects of dilemmas, including the creation of such conditions with the help of various cases and also the rules so that the individual can choose the right alternative and solve the dilemma.

Dilemmas create complexity and conflict in the mind that makes individuals incapable of making quick decisions. Managers have a tough challenge because if employees are not relaxed and satisfied, this can hamper the normal functioning of the organization. Proper codes of conduct should be discussed with the employees; this can give guidelines to the employees whenever they are confronted with any kind of dilemmas. Moreover, providing proper training programs for the employees can also be a solution for managing the dilemma as employees can have brainstorming sessions, and they can develop a proactive approach for evaluating alternatives and the consequences of the decisions.

REFERENCES

Andriopoulos, C., & Lewis, M. W. (2009). Exploitation-exploration tensions and organizational ambidexterity: Managing paradoxes of innovation. *Organization Science, 20*(4), 696–717.

Chin, K.-S., Chan, B. L., & Lam, P.-K. (2008). Identifying and prioritizing critical success factors for coopetition strategy. *Industrial Management & Data Systems, 108*(4), 437–454. 10.1108/02635570810868326

Cuban, L. (1992). Managing dilemmas while building professional communities. *Educational Researcher, 21*(1), 4–11

Farjoun, M. (2010). Beyond dualism: Stability and change as a duality. *Academy of Management Review, 35*(2), 202–225.

Figar, N., & Đorđević, B. (2016). Managing an ethical dilemma. *Economic Themes, 54*(3), 345–362.

Fontin, M. (1997). *Das Management von Dilemmata*. Springer.

Gilbert, G., & Sutherland, M. 2013. The paradox of managing autonomy and control. *South African Journal of Business Management, 44*(1), 15–27.

Hampden-Turner, C. (1990, November). Of tension, harmony & dilemma-busting. Interview by Joe Flower. In *The Healthcare Forum Journal* (Vol. 33, No. 6, pp. 74–78).

Heinze, T., Kizirian, T., & Leese, W. (2004). Fraternization in accounting firms: A case study. *Journal of College Teaching & Learning, 1*(12), 61–64.

Hülsmann, M., & Berry, A. (2004). Strategic Management Dilemma: It's necessity in a World of Diversity and Change. In *Conference Proceedings of SAM and IFSAM VII World Congress: Management in a World of Diversity and Change, Göteborg.*

Johnson, H. T. (2012). Lean dilemma: Choose system principles or management accounting controls—Not both. *Lean Accounting: Best Practices for Sustainable Integration*, 1–16.

Neuberger, B. (1995). National self-determination: Dilemmas of a concept. *Nations and Nationalism, 1*(3), 297–325.

Peters, L. (2012). The rhythm of leading change: Living with paradox. *Journal of Management Inquiry, 21*(4), 405–411.

Rossouw, D., & van Vuuren, L. (2006). *Business Ethics* (3rd ed). Cape Town: Oxford University Press.

Sostrin J. (2015, June 13). Understanding 'The Manager's Dilemma'. https://www.linkedin.com/pulse/understanding-managers-dilemma-jesse-sostrin-phd#:~:text=The%20Manager's%20Dilemma%20is%20a,to%20meet%20them%20do%20not

Steps to Solve an Ethical Dilemma. (2008). Retrieved from: https://biggsuccess.com/2008/08/05/3-steps-to-solve-an-ethical-dilemma/, Accessed on 11 September 2022.

Tracey, P., Creed, W. D., Smith, W. K., Lewis, M. W., Jarzabkowski, P., & Langley, A. (2017). Beyond managerial dilemmas. *The Oxford Handbook of Organizational Paradox, 162.*

Van Gramberg, B., & Teicher, J. (2006). Managing neutrality and impartiality in workplace conflict resolution: The dilemma of the HR manager. *Asia Pacific Journal of Human Resources, 44*(2), 197–210.

Wallace, J. (2015). Dilemmas of science teaching. In R. Gunstone (Ed.), *Encyclopedia of Science Education* (pp. 331–332). Dortrecht, The Netherlands: Springer.

Yin, H. (2015). Dilemmas of teacher development in the context of curricular reform In Q. Gu (Ed.), *The Work and Lives of Teachers in China* (pp. 85–104). London: Routledge.

10 Triple Bottom Line Framework and Sustainable Practices
A Study from the Global Perspective

Tushar Soubhari
Research Scholar, DCMS, University of Calicut, Kerala, India

E K Satheesh
Professor & Research Supervisor, DCMS, University of Calicut, Kerala, India

Sudhansu Sekhar Nanda
Associate Professor, Kirloskar Institute of Management, Harihar, Karnataka

Sasikanta Tripathy
Assistant Professor, College of Business Administration, University of Bahrain, Bahrain

10.1 INTRODUCTION

The term "Triple Bottom Line" (TBL), which John Elkington first used 28 years ago, has been gaining traction as a sustainability-related idea. Studies emphasize clarifying what sustainability actually is, establishing specific, attainable goals, and developing ways to meet those goals. In the middle of the 1990s, he worked to quantify sustainability by including a new accounting framework, which led to the TBL Model. Profits, People, and the Planet are the three Ps most emphasised. The Brundtland Report and George's 1879 concept known as "spaceship Earth" are where sustainability first emerged more than 130 years ago (Brundtland, 1987, p 43). Through its foundation in sustainability, TBL provides a framework for rating an organization's success on three fronts: financial, social, and ecological (Goel, 2010). Although there are differences in how the applications of TBL large corporate houses and non-profit organizations are driven by the concepts of

DOI: 10.1201/9781003365525-10

economic, environmental, and social sustainability. Every businessman feared that during the COVID pandemic's recessionary period, they could turn to economic imperatives rather than TBL. Corporate social responsibility, meantime, grew in popularity as a result of the Internet's availability of empowering knowledge on the shift from a "me" to a "we" society and from ostentatious to conscientious consumption.

Large family-owned businesses in India, including Tata, Birla, Ambani, etc., have historically built hospitals, temples, and schools as investments in the welfare of the local population. Banking institutions in India are engaged in microfinance due to the success of the Grameen Bank in Bangladesh. With the help of various Indian companies, Shell and Pratham have concentrated on primary education. In order to combat malnutrition, Unilever India and Infosys are involved. In addition, ITC provides technology to farmers so they may buy, sell, and harvest their crops with greater knowledge. As a result, social consciousness is considered crucially popularised and becoming a crucial component of today's companies that are revolutionizing sectors and protecting the environment at the same time.

The TBL agenda proposed by Elkington (1997) discusses seven drivers towards sustainability, which is evident from Figure 10.1, shown below:

The paradigm shifts embarking on sustainable capitalism transitions negotiate for business discussions driven by competition; softened values; openness and transparency in communication; process-based life cycle; symbiotic partnerships; and inclusive governance over a longer time period. It is evident from Figure 10.2, shown below:

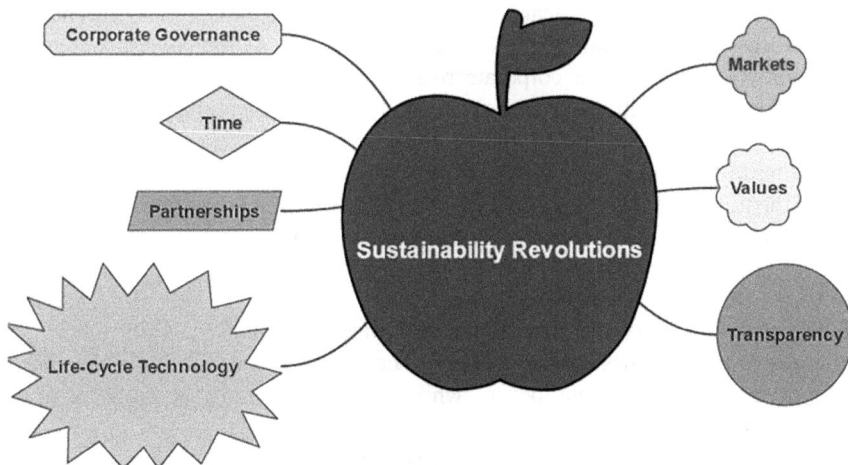

FIGURE 10.1 Showing seven drivers of sustainability.

Source: Seven Sustainability Revolutions (Elkington, 1997) modified for the study.

Old Paradigm		New Paradigm	Drivers
Compliance	→	Competition	Market
Hard	→	Soft	Values
Closed	→	Open	Transparency
Product	→	Function	Life-Cycle Technology
Subversion	→	Symbiosis	Partnerships
Wider	→	Longer	Time
Exclusive	→	Inclusive	Corporate Governance

FIGURE 10.2   Showing paradigm shifts in sustainability revolutions.

Source: Paradigm shifts in Sustainability Revolutions (Elkington, 1997) modified for the study.

10.1.1   THREE PRESSURE WAVES WITNESSED BY ELKINGTON SINCE 1997

Since 1960, the environmental agenda has been moulded by three distinct waves of popular pressure. Governments and the public sector have adapted their roles and responsibilities in response to each of these three waves.

- *WAVE 1* increased awareness of the need to limit environmental impacts and demands on natural resources, which led to the first wave of environmental legislation. At its best, the corporate response was defensive and compliance-focused.
- *WAVE 2* increased awareness that new product types and production methods may be required, leading to the need for sustainable development processes. The response from the business world started to heat up.
- *WAVE 3* emphasized the growing recognition that sustainable development would necessitate significant reforms to corporate governance and the entire process of globalization, putting the spotlight back on the government and non-profits. Now, the corporate response would need to concentrate on market creation in addition to the regulatory and competitive dimensions.

According to Elkington (1997), the secret to creating environmental regulations is to ease the transition towards sustainability and recognize that the four various types of organizations require diverse roles from the government.

10.1.1.1   Corporate Locusts

Many businesses carry out detrimental activities continuously, while others just periodically act in a destructive manner. The social and environmental value is being destroyed by corporate locusts, who are also weakening the basis for future economic progress.

The following are some additional characteristics of a corporate locust, according to Elkington (1997):

- ecological, human, social, and economic capital destruction;
- a business model that can't last;

• and swarming tendencies (think gold rushes), which exceed the carrying capacity of the ecosystem.

10.1.1.2   Corporate Caterpillars

Caterpillars typically are more difficult to see than locusts because of their more restricted effects. The degenerative effects of a corporate caterpillar may make it difficult to realize that these businesses have a sizable potential for transformation if we live or work directly next to one.

Corporate caterpillars typically:

• show laser-like concentration on the immediate goals of the company;
• rely on a high "burn rate," yet typically with renewable forms of financing;
• rely on an economic model that falls apart when scaled up to a more equitable world of 8–10 billion people.
• build on proven business principles and can evolve into a more environmentally friendly form.

Elkington (1997), however, assessed the government's efforts and suggested a number of improvements:

• green purchasing,
• public–private partnerships,
• and reforms to the ecological tax system are all examples of how we may help the environment while also improving the economy.
• the end of inefficient subsidies

10.1.1.3   Corporate Butterflies

Despite the fact that the majority are rather little, corporate butterflies are simple to recognize. They are frequently highly noticeable by nature, and in recent years, the media has covered them extensively. They serve as models for honeybees to imitate and, most importantly, scale up new strategies for generating sustainable prosperity.

Their key characteristics are:

• Inclusion in the Global Competitiveness Report;
• company-wide dedication to CSR and SD initiatives;
• a propensity for articulating its place in terms of insects like caterpillars and locusts;
• a vast network, although not among honeybees or caterpillars;
• and a financially viable business model, though the latter may become less so as success fuels growth, expansion, and an ever-increasing reliance on financial markets and large corporate partners.

10.1.1.4   Corporate Honeybees

In the ensuing decades, an increasing number of governmental organizations, inventors, businesspeople, and investors will turn to this field. A thriving global economy buzzes with the efforts of corporate honeybees and their colonies.

	Low Impact	High Impact
Regenerative (increasing returns)		
Degenerative (decreasing returns)		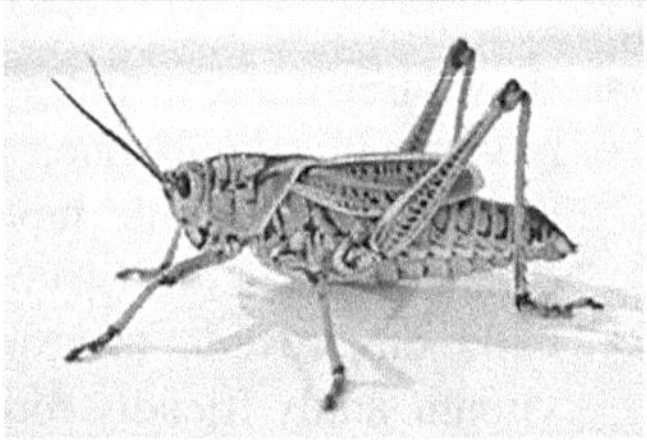

FIGURE 10.3 Showing corporate characteristics based on animal spirits.

Source: Corporate Characteristics (Elkington, 1997) modified for the study.

Although bees may swarm frequently, much like locusts, their impact is long-lasting and immensely restorative.

Elkington (1997) identified the following as the essential traits of the corporate honeybee:

- the ability to sustainably produce natural, human, social, institutional, and financial resources;
- a method of doing business that relies on ongoing innovation;
- an established and applicable code of corporate ethics;
- the ability to socialize and evolve powerful symbiotic partnerships.

The above explanations have been precisely summarized in the form of Figure 10.3, as shown above.

10.2 LITERATURE REVIEW

10.2.1 TRIPLE BOTTOM LINE (TBL)

Elkington coined the TBL framework, which has been called "a smart and far-reaching metaphor". Before the late 1990s, the expression was largely unknown. By bringing together the economic and social dimensions, TBL demonstrates how the environmental agenda can be expanded (Elkington, 1997). It provides a framework for assessing an organization's impact across economic, social, and environmental dimensions (Goel, 2010). Another name for this concept is the sustainable development framework. Targeted at businesses, the TBL agenda uniformly and equitably

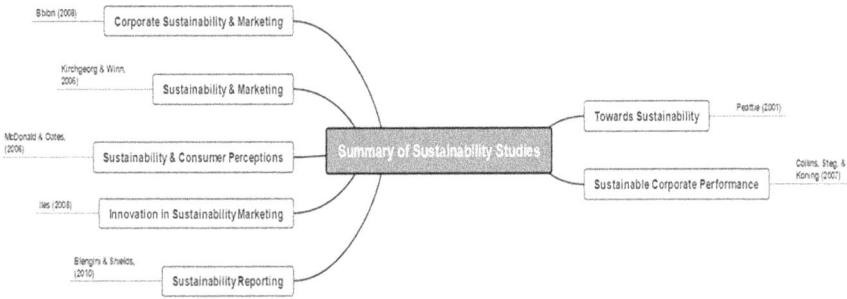

FIGURE 10.4   Showing reviews of previous literature works on sustainability.

Source: Created by the Authors using EdrawMind Wondershare tool software ©.

highlights the contributions of firms to economic growth, social welfare, and environmental protection.

When discussing the TBL framework, the term "economic line" is used to describe the influence of the business's operations on the economy as a whole (Elkington, 1997). As a part of the sustainability framework, it deals with the economy's ability to persist and grow in the future, so that it can provide for future generations. The organization's ability to promote economic growth is represented by the slope of the economic line, which is a function of how closely the organization's growth tracks with economic growth. As such, it highlights the ways in which the company improves the environment economically, hence increasing its viability for future generations.

The social line of TBL is exemplified by a company's commitment to and implementation of socially responsible and mutually beneficial community, labour, and human capital management practices (Elkington, 1997). The idea is to help the community and "give back" to those who have helped you.

The TBL's environmental line emphasizes safeguarding the planet's natural habitats and resources for the benefit of future generations. Reduced energy consumption, reduced emissions of greenhouse gases, reduced ecological footprints, etc. are all aspects of this concept (Goel, 2010). The social aspect of TBL, like environmental activities, affects the long-term viability of businesses. The financial benefit is the result of lower operational costs and higher profits from the development of innovative green products (Kearney, 2009).

10.3   3 P FORMULA FOR SUSTAINABILITY

The purpose of the 3Ps in sustainable development is to secure and maximize their advantages. Sustainability projects prioritize humanity. By respecting the planet's precious resources and seeking to make a profit without wasting anything, they acknowledge the requirements of the planet.

10.3.1   PROFITS

The bottom line or the amount of money returned to investors, is the ultimate yardstick of corporate success. The key choices and planning activities of a corporation are meticulously developed to maximize profits while minimizing costs and risks. Leaders with a sense of purpose know they can make a positive difference in the world through their businesses without compromising on profits.

10.3.2 PEOPLE

In order to maximize their own value, large firms have traditionally focused on increasing their stockholders' wealth. More and more companies are realizing the importance of incorporating sustainability into their operations, and as a result, they are shifting their focus to creating value for all stakeholders (customers, employees, and the local community) through fruitful strategic alliances with non-profit organizations that have a similar mission.

10.3.3 PLANET

Since the beginning of the Industrial Revolution, big businesses have produced an incredible amount of environmental pollution, a key contributor to climate change. The latest data from the Carbon Majors Database shows that only 100 energy-related corporations are responsible for roughly 71% of all industrial emissions. But historically, companies have been the primary contributors to global warming. They have begun making substantial contributions to society, the environment, and the economy after realizing their mistakes and assuming responsibility for these issues (Figure 10.5).

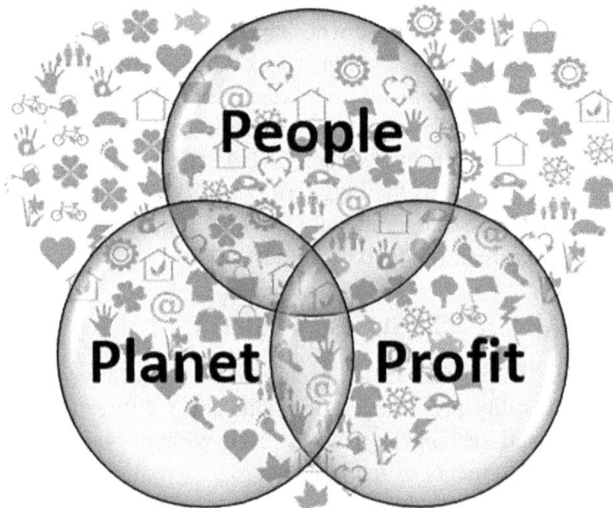

FIGURE 10.5   Showing the 3 P's in sustainability (outlined under the Triple Bottom Line).

Source: Modified by the Authors based on works of Elkington, 1997.

10.3.3.1 Triple Bottom Line Business Operations in India

In 2021, 80% of the top 100 companies in terms of sustainability and CSR will incorporate Sustainable Development Goals (SDGs) into their responsible business practises by 2030. The top 25 firms on Futurescape's ranking of the best for sustainability and CSR in 2021 also map their corporate objectives to the SDGs.

The table below summarizes several conclusions about their practices based on the sustainable and TBL operations of top leading firms in India:

Company Name	Profit Orientation	People Orientation	Planet Orientation
Godrej Consumer Products Ltd.	Funded CSR projects to the tune of Rs. 34.08 cr. in FY 2020–21.	Helps over 9000 people who own Nano businesses by implementing medium- to long-term livelihood rehabilitation programmes. Approximately 2.77 million people from low-income and other marginalized communities benefited from the company's CSR programmes.	During the previous fiscal year, we were able to achieve water positivity and send no trash to landfills. The fulfilment of all of Extended Producer Responsibility's requirements.
Infosys Ltd.	Invested Rs 325.32 billion in community service projects	Sustainable development, institution-building, and community-wide improvement projects. Infosys' ESG Vision 2030 called for the rollout of a comprehensive digital skills programme, and so the company launched Infosys Headstart. By 2025, more than 10 million people will have been trained in digital and life skills.	Putting an emphasis on environmentally responsible and socially just business practises.
Wipro Ltd.	The amount allotted for CSR was Rs 251 billion.	Has funded over 1,561 projects that have helped over 10 million people as part of its COVID-19 response. These projects include those that provide humanitarian aid, integrated healthcare support, and livelihood regeneration. contributed to the economic recovery of over 8.2 million individuals and the operations of over 500 charities that deliver humanitarian and medical help.	Product sustainability and effective electronic waste recycling.

(Continued)

Company Name	Profit Orientation	People Orientation	Planet Orientation
Tata Chemicals Limited	CSR spending for the 21st fiscal year came to Rs. 21 crores.	Over the course of 6878 online and face-to-face interactions, farmers received assistance with livestock management, field demonstrations, and skill development. Okhai's rural women artisans were given a helping hand, and the region was transformed into a bazaar that attracts customers and 25,190 artists from all over India.	In order to preserve Mithapur's native flora, the city's foremost sustainability champions planted 1.15 million mangrove trees in a wide variety of sites.
ITC Ltd.	In the 2020–21 financial year, the corporation budgeted Rs 353.46 crores for CSR initiatives.	During the year, 640 individual household toilets were erected in 28 districts, while vocational training programmes helped 12,470 youth.	It is estimated that the Social Forestry Programme has improved the quality of life on 30,439 acres of land. About 70,900 metric tonnes (MT) of dry waste were collected from 1,067 wards by ITC's "Well Being Out of Trash (WOW)" waste recycling project.
Jubilant Life Sciences Ltd.	Invested Rs 5.83 billion in community service initiatives.	Nearly 6.5 million people in 240 communities have benefited from these initiatives. These programmes focus on health, education, sustainable livelihoods, and social entrepreneurship. Initiatives to improve livelihoods include Samridhi (Self-Help Group and Micro-Enterprise Promotion) and Jubifarm (Sustainable Agriculture Program).	Uses the Jubilant Bhartia Foundation as the vehicle for most of its corporate social responsibility activities.
Hindustan Lever Ltd. (HUL)	Make a monetary commitment of Rs. 100 crores ($1 billion) to social causes in order to supplement the	Through a collaboration with the government and other NGOs, over 2 crores' worth of supplies were donated to frontline medical professionals, police officials,	Its "Water for Public Good" programme was founded on the principle that water resources should be controlled at the community level.

Company Name	Profit Orientation	People Orientation	Planet Orientation
	government's efforts in dealing with Covid-19.	sanitation workers, and vulnerable inhabitants of the country. They collaborated with healthcare providers to provide 75,000 test kits to the public. Near our facilities and offices, public health officials from many states worked together on this initiative. Collaborated with the State Bank of India, OYO, Lemon Tree, and Apollo Hospitals to set up isolation wards with medical monitoring in major cities to reduce stress on existing medical facilities. The company and the government in Haridwar and Nasik also opened an isolation hospital with 30 beds in the previous year.	keeping up the call for people to embrace three healthy habits (or "Swachh Aadat"): washing hands five times a day, defecating in designated areas, and utilizing safe water practices.
Apollo Tyres	Invested Rs. 12,9 billion.	Offers care for a wide range of conditions, including HIV/AIDS prevention and education, eye care, tuberculosis integration, and the management of non-communicable diseases like diabetes and hypertension.	Used tires served as the initial inspiration for the ELT Playgrounds initiative. Tire-based play structures help kids get in shape and fit while also teaching them about recycling and reusing materials. Due to the establishment of another ELT playground in the fiscal year 2021 at the Karanja, Maharashtra site, there are now a total of 14 such playgrounds around the country.
Hero Motor Corp.	CSR spending for 2020–21 was Rs 99.73 cr.	A total of 23 million meals, 34,000 ration packages, 37,201 litres of hand sanitiser, 44,000 masks, and 44,000 personal protective equipment kits were distributed to government	During the lockdown, a small number of motorbikes were modified into "First Responder Vehicles" (FRVs) to help the medical community and mobilize staff.

(Continued)

Company Name	Profit Orientation	People Orientation	Planet Orientation
		hospitals, police departments, and other agencies to combat the spread of COVID-19.	Sixty FRVs, all different makes and models, were donated to the government for use as ambulances.
Maruti Suzuki India Ltd.	A total of Rs 140.940 Cr was allocated to CSR projects during the 2020–21 fiscal year.	Community development activities focusing on water, sanitation, health, education, and shared community resources were prioritized in the 26 villages surrounding its facilities in Gujarat and Haryana.	Potable water ATMs, water supply infrastructure, individual toilet construction, and sewer line development.

10.3.3.2 Triple Bottom Line Practices from a Global Perspective

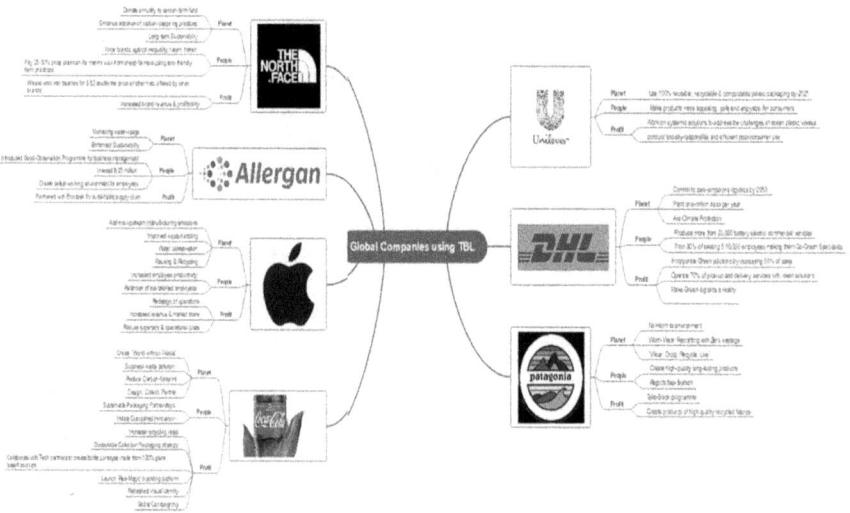

FIGURE 10.6 Showing the bird's eye view of Global Companies' TBL Strategies.

Source: Designed by the Authors based on case histories of Global Companies on TBL Strategies.

10.4 IMPLICATIONS OF THE STUDY

The study discussed here included different cases from both India and globally, encrypting various reasons for how they move forward the Triple Bottom Line framework and sustainable practices as a lucrative strategy in the modern era. The findings from various studies summarize the fact that such companies raise transparency by mitigating their shareholders' concerns of concealed information; involving accountability of their day-to-day actions; delivering growth; and maintaining economic stability for the company itself. Moreover, these corporates have managed to give a competitive advantage over their industry peers by strengthening their business objectives, paving ways to betterment at the global level, minimizing risks of public scrutiny, and propelling our world to be made a better place to live in. Hence, sustainability stands to be the benchmark for our future in the coming years. The study concludes at this point.

REFERENCES

Brundtland, G. (1987). *Our Common Future: The World Commission on Environment and Development*. Oxford, England: Oxford University Press.

Elkington, J. (1994). Towards the sustainable corporation: Win-Win-Win business strategies for sustainable development. *California Management Review* 36(2), 90–100.

Elkington, J. (1997). *Cannibals With Forks – Triple Bottom Line of 21st Century Business*. Stoney Creek, CT: New Society Publishers.

Goel, P. (2010). Triple bottom line reporting: An analytical approach for corporate sustainability. *Journal of Finance, Accounting, and Management*, 1(1), 27–42.

Kearney, A. (2009). "Green" winners: The performance of sustainability-focused organizations during the financial crisis. Retrieved September 12, 2009 from http://www.sustaincommworld.com/pdfs/ATKearney_Green_Winners.pdf

WEBSITES

https://thecsrjournal.in/top-100-companies-india-csr-sustainability-2021/

https://www.sentinelassam.com/editorial/the-triple-bottom-line-485671

https://earth911.com/business-policy/triple-bottom-line-7-companies/

https://buzzonearth.com/blog/2020/10/05/most-responsible-indian-companies-in-terms-of-sustainability-and-csr/

https://www.business-standard.com/article/economy-policy/the-power-of-the-triple-bottom-line-107061501011_1.html

https://www.ibrc.indiana.edu/ibr/2011/spring/article2.html

https://economictimes.indiatimes.com/blogs/Globalpositioning/the-triple-bottomline/

https://online.hbs.edu/blog/post/what-is-the-triple-bottom-line

https://youtu.be/1-Ct_53XKYY

https://www.johnelkington.com/archive/TBL-elkington-chapter.pdf

https://bstrategyhub.com/what-is-triple-bottom-line-tbl-explained-with-examples-the-future-benchmark/#3_Patagonia_-_reuse_old_clothes

https://www.greenbusinessmba.com/blog/what-is-the-triple-bottom-line

https://www.ibrc.indiana.edu/ibr/2011/spring/article2.html

https://embapro.com/frontpage/tblcoanalysis/7669-apple

11 Impact of Advertising on Shopping Behaviour
A Study of Mobile Phones

Sapna Yadav
Jaypee Business School, JIIT, Uttar Pradesh

Meenakshi Tyagi
Associate Professor, KIET Group of Institutions, Delhi- NCR, Ghaziabad

Himani Grewal
Assistant Professor, Moradabad Institute of Technology, Moradabad

Sonia Gouri and Tanushree Sanwal
Assistant Professor, KIET Group of Institutions, Delhi- NCR, Ghaziabad

11.1 INTRODUCTION

Nowadays advertisements attract buyers to commercials and control their public behaviour, purchasing behaviour, and their mindset. The revelation of so many advertisements converts human desires into essentials and makes them purchase items that should not be their preference over necessary items. Advertising affects consumers' choices in switching brands and changing their loyalty with brands. The present study intends to highlight the impacts of commercials in triggering the potential buyers for buying mobile phones. Only branded mobile phones are believed worth buying while the market has a significant range of local mobile phones. Therefore, the present research focuses on the study of the consequences of commercials on buyers' purchasing conduct with reference to smart phones by examining the level of advertisement influence in purchasing behaviour, effectiveness of advertisements in smart phone purchase decisions, awareness, and purchasing behaviour of customers.

In the present time of technology and globalization, an efficient commercial is significant to affect buyers' purchasing conduct in the mobile phone sector. In this research, we also focus on how the mobile phone advertisement compels the

DOI: 10.1201/9781003365525-11

consumer to switch their brand affiliation. The swift increase of mobile phones and other related devices has established a new path for marketing. There are many features in mobile phones like texting, video and voice calls, e-mailing, etc., that enable users to be online anywhere for required usage. The usage of a mobile device is very high. That is why the mobile commercial reaches more potential buyers.

11.2   LITERATURE REVIEW

Presently social networking has increased the communication of marketers and customers because of its dynamic nature. This has developed the way of advertising products and communicating with customers. In modern times, Facebook is one of the most used media platforms, which covers huge instantaneous responses from its network users for framing and trending a perception [1,2]. In fact, this strong social media platform has built an excellent opportunity for any product or brand to expose its features through publicity, awareness, and insight; to frame perception; and to set standards [3]. The practice of customary one-way communication to encourage customer opinion and enhance positive mindsets for product value has been significantly dropping its convincing impact due to the all-encompassing influence of Facebook as a system of linking people [1,3]. The processes of product estimation, perception, and attitude development have now been strongly linked to a new multidimensional communications system where customers are more involved and find more reliability through following and restructuring people's views rather than receiving old-style advertising [4–7].

The fastest-growing businesses are pushing hard to support product consideration and exposure to build a constructive reputation through viral marketing on social networking sites; these businesses now recognize that the social media space is the key hub for spreading the early word of a product's launch and inspiring people to use it [8]. Nearly 60% of Facebook's 1.6 billion active users watch advertising for products, making up the majority of the social network's user base. Kim and Ko estimated that nowadays, 70% of active social network users learn more about products before making an actual purchase and get various ideas about shopping through social media sites. More than one billion users who are connected to Facebook have access to a fantastic platform that businesses can use to increase their product advertising.

Any type of product detail's impact is greatly influenced by its source [7,9,10]. Marketing sources through reliability, consistency, steadiness, and pursuit of value in cultivating a positive mindset, create a significant potential impact on customers [7,9]. Companies are showing great interest in the modern way of marketing and using internet marketing in place of the traditional manner of marketing. That is why the use of social networks has grown [10]. Facebook has become one of the most liked social sites for online marketing that enables constant two-way interaction without location and time constraints, and it is seen as a trustworthy and universal alternate route for information distribution as well [9].

Businesses benefit from e-commerce and online purchasing in terms of income, which is further enhanced by employees' high levels of social intelligence [11,12].

11.3   OBJECTIVES OF THE STUDY

1. To study the level of impact of mobile phone advertising on consumer buying behaviour
2. To know whether these advertisements provide some useful information or just manipulating
3. To analyze the consumer's expectation from an advertisement.

11.4   HYPOTHESIS/NULL HYPOTHESIS

H_{o1}: There is no difference in the satisfaction of male and female students.
H_{o2}: Customer satisfaction is not impacted according to their locality.
H_{o3}: Customer satisfaction is the same for customers of different age groups.

Alternate Hypothesis

H_{11}: There is a difference in the satisfaction of male and female students.
H_{12}: Customer satisfaction is varied according to their locality.
H_{13}: Customer satisfaction is different for customers of different age groups.

11.5   RESEARCH METHODOLOGY

Primary data in the present study were collected from the students of five different colleges of Delhi-NCR through a questionnaire. The collected data were analyzed through statistical tools for fulfilment of the objectives.

Sources of data: Sources of data are primary and secondary data.

Primary data: A well-framed questionnaire is prepared and used for the collection of primary data.

Secondary data: Secondary data is gathered from published sources like standard text and reference books, journals, magazines, and the internet, which is related to the study.

SAMPLE DESIGN

Random sampling was used to identify the sample respondents from a total population of 215 students. These respondents were selected from different departments of different institutions. For collection of the true and accurate information, students have been called and explained the research problem, and their consent has been received for responding to the questionnaire.

TOOLS AND TECHNIQUES: For analyzing the collected data simple, percentage analysis has been used in this study, and to test the hypothesis, t-test and F-test were used.

11.6   DATA ANALYSIS AND INTERPRETATION

In these data, we have noticed that around 125 people came across the advertisement through the internet; 22 people came across the advertisement through a magazine;

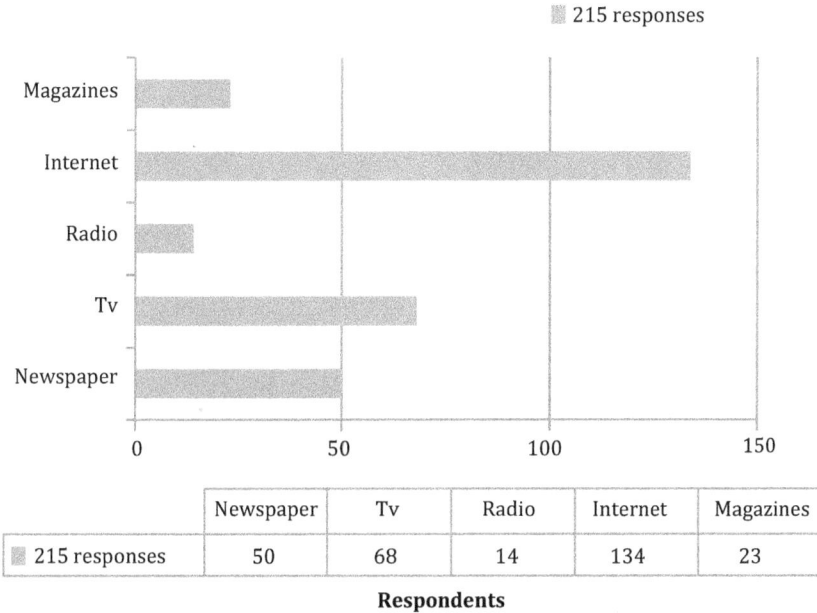

	Newspaper	Tv	Radio	Internet	Magazines
215 responses	50	68	14	134	23

Respondents

FIGURE 11.1 Medium through which consumer come across the advertisement related to mobile phones.

17 people through the radio, 65 people through TV; and 50 people through the newspaper (Figure 11.1).

In Figure 11.2, it is clear that 47% of the individuals came in contact with advertisements every day; 23% of individuals came in contact a couple of times a week; 16% came in contact once a week; 10% came in contact once a month; and only 4% came in contact once a year.

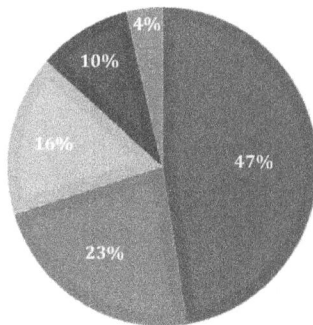

FIGURE 11.2 How often do you come in contact with advertisements targeting you directly?

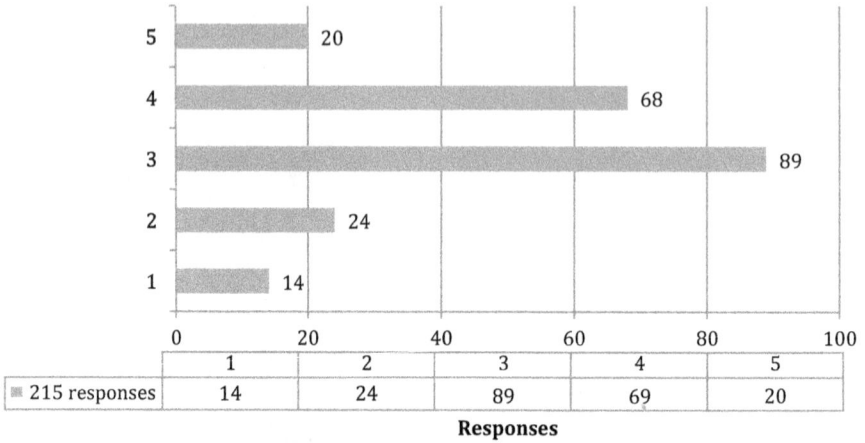

▦ 215 responses	1	2	3	4	5
	14	24	89	69	20

Responses

FIGURE 11.3 Degree of confidence in these advertisements (1 shows the least while 5 is high).

From Figure 11.3, it is clear that out of 215 respondents, 89 respondents had a moderate level of confidence in these advertisements, and a further 68 respondents had a highly moderate level of confidence. Similarly, only 14 respondents had a low level of confidence in these advertisements.

It is clear from Figure 11.4 that 41.8% individuals are making their own decision while purchasing a mobile phone. Friends also play an important role in decision making while purchasing a mobile phone. Out of 215 people, 47 individuals made decisions after discussing with friends.

From Figure 11.5, 31% of respondents have knowledge of mobile phone applications; 18% of respondents have very good knowledge about mobile phone applications; and 25% of respondents have a good knowledge about mobile phone applications. A total of 26% of respondents had limited or very limited knowledge about mobile phone applications.

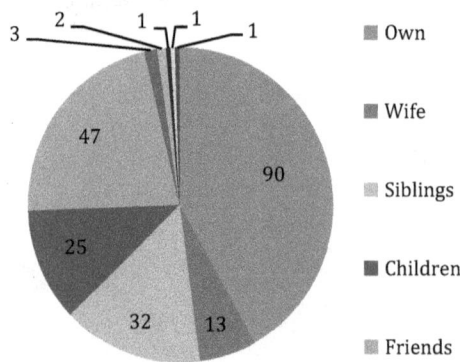

FIGURE 11.4 Decision on purchase.

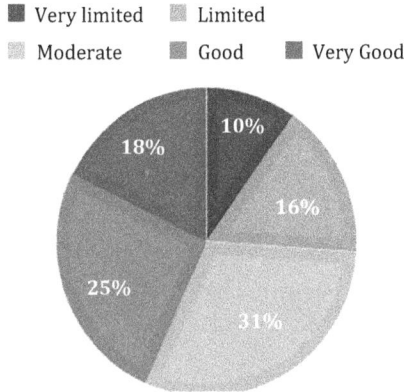

FIGURE 11.5 Knowledge about mobile phones.

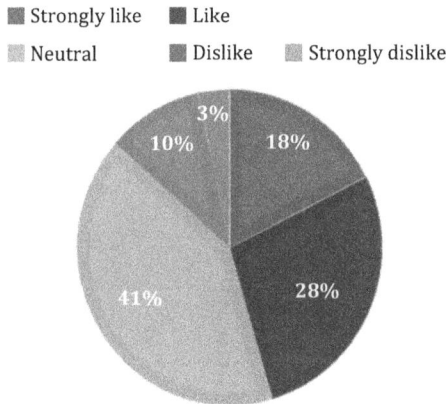

FIGURE 11.6 Attitude towards the social media advertisements.

From Figure 11.6, it is clear that 41% of individuals are neutral towards social media advertisement and 28% like social media advertisements. Only 3% strongly dislike social media advertisements.

Figure 11.7 gives an overview that a 1 rating represents the least and a 5 rating is the highest. The individuals who gave a 1 rating believe that the information in the advertisement does not affect their buying decision, whereas those people who gave a 5 rating strongly believe that the information in the advertisement content affects the buying decision. Ninety-two individuals rated the maximum out of the total of 215 respondents.

It is clear from Figure 11.8 that 37% of individuals are neutral in the purchase decision through TV and newspaper advertisements, while 31% of individuals agreed while making decisions through TV and newspaper advertisements. Only 5% of individuals strongly disagreed with the same.

A 1 rating is the least, whereas a 5 rating is the highest. A person who gave a 1 rating means they do not rely on the internet advertisement, whereas those who

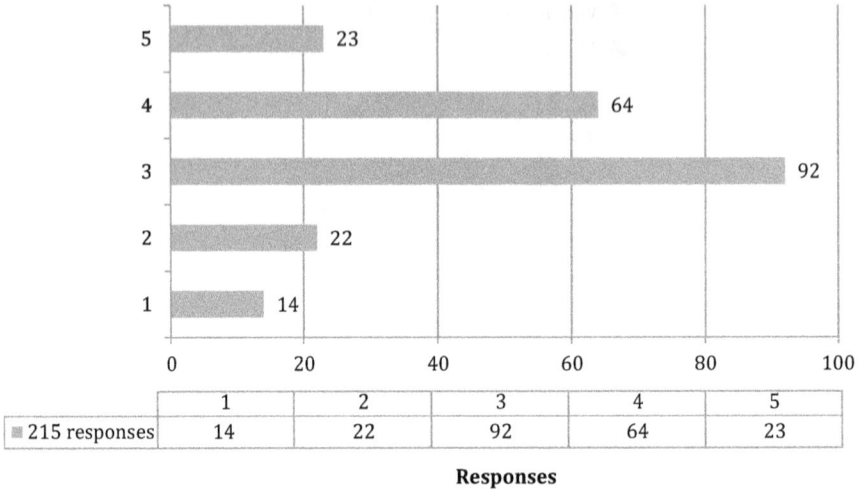

■ 215 responses	1	2	3	4	5
	14	22	92	64	23

Responses

FIGURE 11.7 Information in the advertisement content affects buying decision.

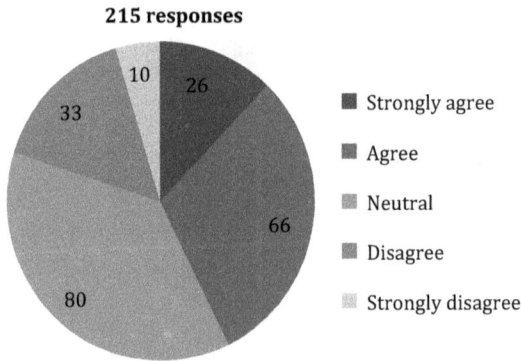

FIGURE 11.8 Purchase decision through TV and newspaper advertisements.

gave a 5 rating rely more on the internet advertisement while buying a new phone. From Figure 11.9, it is clear that 77 people rely on the internet advertisements while buying a new phone, and only 14 people did not rely on the same.

From Figure 11.10, it is clear that the majority of people think that Xiaomi (Redmi) company invests a maximum amount, i.e., 34%, in the advertisement of its products, followed by Samsung company, which is 28%, and only 1% invested in One Plus company.

From Figure 11.11, it is clear that 58% of people purchased a phone after considering these advertisements, and 42% did not purchase it after considering these advertisements.

From Figure 11.12, it is observed that 41% of people strongly believed that the battery life was exactly as shown in the advertisement, while 29% had a neutral opinion. Only 4% of people strongly disliked that the battery life was exactly as shown in the advertisement.

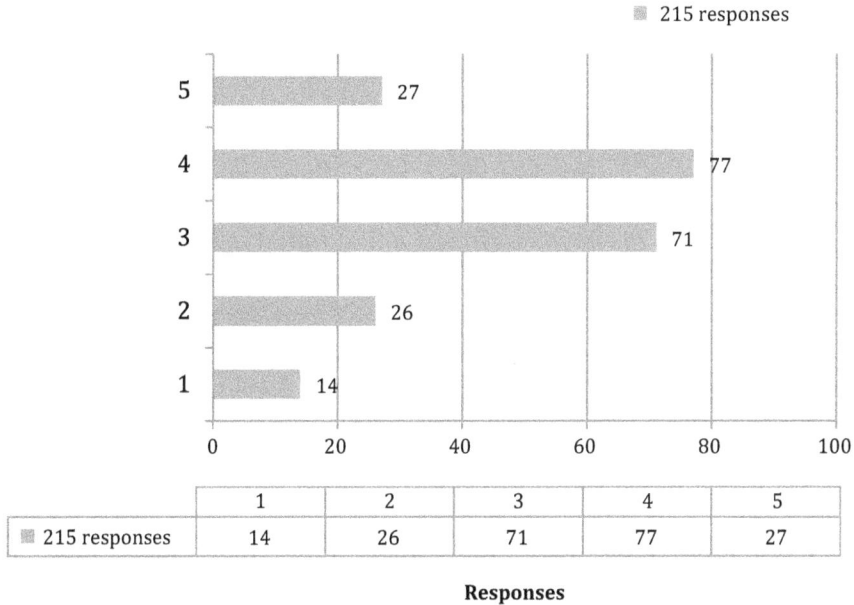

	1	2	3	4	5
215 responses	14	26	71	77	27

Responses

FIGURE 11.9 Extent to which people rely on the internet advertisements while buying a new phone.

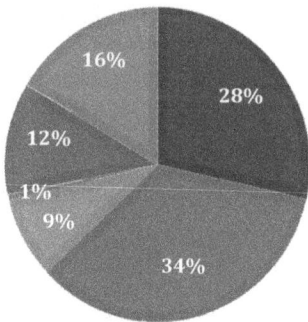

FIGURE 11.10 Mobile company invest maximum amount in the advertisement of its products.

Figure 11.13 indicates that 42% of people liked that the camera quality is exactly as shown in the advertisement, followed by 24% of people who were neutral, 21% who strongly agreed, and very few, i.e., 4% of people, who strongly disliked that the camera quality was exactly as shown in the advertisement.

Figure 11.14 indicates that 44% of people liked that the display size that is mentioned in the advertisement matches the actual size, followed by 27% who strongly liked it, 22% were neutral, and very few (i.e., 6% people) disliked or strongly disliked when the display size that is mentioned in the advertisement matches with the actual size.

Yes   No

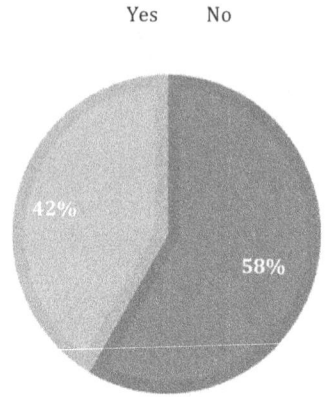

FIGURE 11.11 Any phone bought after considering these advertisements.

■ Strongly Like ■ Like
■ Neutral ■ Dislike ■ Strongly Dislike

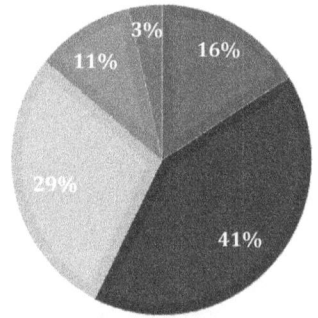

FIGURE 11.12 Battery life same as shown in the advertisement.

■ Strongly Like ■ Like
■ Neutral ■ Dislike ■ Strongly Dislike

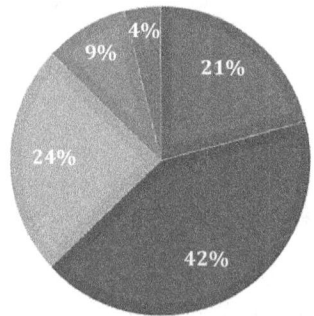

FIGURE 11.13 Camera qualities that are mentioned in the advertisements are true.

■ Strongly Like   ■ Like
■ Neutral   ■ Dislike   ■ Strongly Dislike

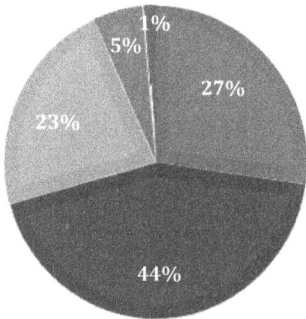

FIGURE 11.14   Display size that is mentioned in the advertisement matches with actual size.

11.7   HYPOTHESIS TESTING

Null Hypothesis H_{o1}: There is no difference in the satisfaction of male and female students.
Alternative Hypothesis H_{11}: There is a difference in the satisfaction of male and female students.

The mean value for customer satisfaction for male and female students is 20.71 and 21.40, whereas the standard deviation is 1.25 and 1.55 (Table 11.1). To test the difference statistically, we have tested an independent sample t-test. The significant value is 0.23 (Table 11.2), which is higher than .05, which means that there is no difference in the satisfaction of male and female customers.

TABLE 11.1
Gender Statistics

	Gender	N	Mean	Std. Deviation
Satisfaction	Male	135	20.7185	1.25
	Female	80	21.4	1.55

TABLE 11.2
Independent Samples Test

Independent Samples Test

		t-test for Equality of Means		
		T	df	Sig. (2-tailed)
Satisfaction	Equal variances assumed	−1.204	213	0.23
	Equal variances not assumed	−1.26	189.396	0.209

TABLE 11.3
Independent Samples Test

	Sum of Squares	Mean Square	F	Sig.
Between Groups	174.98	58.327	3.756	0.012
Within Groups	3276.852	15.53		
Total	3451.833			

TABLE 11.4
Independent Samples Test

ANOVA					
	Sum of Squares	df	Mean Square	F	Sig.
Between Groups	249.462	3	83.154	5.479	0.001
Within Groups	3202.37	211	15.177		
Total	3451.833	214			

Alternative Hypothesis H_{12}: Customer satisfaction is varied according to different localities.
Null Hypothesis H_{o2}: Customer satisfaction is not impacted according to different localities.

The F-static is 3.756 (Table 11.3), which is significant at .012 for customer satisfaction according to different occupations. This means that customer satisfaction is varied according to different localities. Residence area (urban, semi-urban, rural, semi-rural) has an effect on the satisfaction of the customer.

Null Hypothesis H_{o3} Customer satisfaction is the same for customers of different age groups.
Alternative Hypothesis H_{13}: Customer satisfaction is different for the customers of different age groups.

The F-statics is 5.479, which is significant at .001 for customer satisfaction according to different age groups. This means that customer satisfaction is different for customers of different age groups (Table 11.4).

11.8 CONCLUSION

It has been observed by analyzing all the factors that impact consumer behaviour that one of the major sources to create sensations in customers and encourage them to purchase the products of an advertized mobile phone is effective commercials. Analysis indicates that feelings of pleasure and confidence were created in buyers after watching the commercials as an emotional response for that mobile phone.

High internal consistency was found in all the items in the reliability analysis, which indicates the authenticity of responses for this study. A critical relationship is found between commercials and customer's buying behaviour. Findings about pleasure indicate that potential/buyers feel happy, pleased, hopeful, satisfied, as well as relaxed. That is why findings about arousal indicate the customer's delight and satisfaction. It has been viewed that ads lead to change in preference of the customer, and the customer's thinking matches with the advertisement message. Therefore, it is understood that the advertisement is likable, engaging, and attracts the customer's consideration. It is concluded from outcomes that customers feel good about the display of information presented in commercials through pictures and taglines. Customers are gratified with that information rate because advertisements correctly informed them what they were demanding to know. It has been found from the study that age and residential location has an effect on customer satisfaction, while gender has been found insignificant.

Around 48% of people have a high degree of dependency on these ads while making a purchase decision, and the majority of the respondents have bought a phone after considering these ads.

Around 60% of respondents agree to the features that are shown in advertisements are exact. Adults (20–30 yrs. of age) wish to buy expensive mobile phones, whereas their parents and old age people don't.

Consequently, buyers think that commercials about Redmi (Xiaomi) are more appealing, so results showed that people prefer to purchase them in comparison to other cell phones, and these mobiles are becoming popular due to their features and cost-effective quality.

11.9 LIMITATIONS

- There are many variables that have an impact on consumer's purchasing behaviour and can influence their preference for mobile phones like their experience from past purchases, budget constraints, family pressure, their psychology, etc.
- Effective commercial is not the only source to affect the buyers.
- In the present study, data have been collected from various colleges of Delhi NCR, from candidates of different age groups, so this can also have a different impact on this research study.
- There is a lack of genuine responses.

REFERENCES

1. E. Akar and B. Topçu, "An examination of the factors influencing consumers' attitudes toward social media marketing," *J. Internet Commer.*, vol. 10, no. 1, pp. 35–67, 2011, doi: 10.1080/15332861.2011.558456.
2. A. J. Kim and E. Ko, "Do social media marketing activities enhance customer equity? An empirical study of luxury fashion brand," *J. Bus. Res.*, vol. 65, no. 10, pp. 1480–1486, 2012, doi: 10.1016/j.jbusres.2011.10.014.

3. A. J. Kim and E. Ko, "Impacts of luxury fashion brand's social media marketing on customer relationship and purchase intention," *J. Glob. Fash. Mark.*, vol. 1, no. 3, pp. 164–171, 2010, doi: 10.1080/20932685.2010.10593068.

4. R. Algharabat, N. P. Rana, Y. K. Dwivedi, A. A. Alalwan, and Z. Qasem, "The effect of telepresence, social presence and involvement on consumer brand engagement: An empirical study of non-profit organizations," *J. Retail. Consum. Serv.*, vol. 40, no. September 2017, pp. 139–149, 2018, doi: 10.1016/j.jretconser.2017.09.011.

5. R. Aswani, A. K. Kar, P. V. Ilavarasan, and Y. K. Dwivedi, "Search engine marketing is not all gold: Insights from Twitter and SEOClerks," *Int. J. Inf. Manage.*, vol. 38, no. 1, pp. 107–116, 2018, doi: 10.1016/j.ijinfomgt.2017.07.005.

6. J. L. Hayes and K. W. King, "The social exchange of viral ads: Referral and coreferral of ads among college students," *J. Interact. Advert.*, vol. 14, no. 2, pp. 98–109, 2014, doi: 10.1080/15252019.2014.942473.

7. J. Lu, J. E. Yao, and C. S. Yu, "Personal innovativeness, social influences and adoption of wireless Internet services via mobile technology," *J. Strateg. Inf. Syst.*, vol. 14, no. 3, pp. 245–268, 2005, doi: 10.1016/j.jsis.2005.07.003.

8. N. G. Barnes, P. D, and E. Mattson, "Social media in the 2009 Inc. 500: New tools & new trends conducted by," *Media*, 2009.

9. S.-C. Chu, "Viral advertising in social media," *J. Interact. Advert.*, vol. 12, no. 1, pp. 30– 43, 2011, doi: 10.1080/15252019.2011.10722189.

10. E. L. Pelling and K. M. White, "The theory of planned behavior applied to young people's use of social networking web sites," *Cyberpsychology Behav.*, vol. 12, no. 6, pp. 755–759, 2009, doi: 10.1089/cpb.2009.0109.

11. Sanwal, T., & Sareen, P. (2022). Higher employee engagement through social intelligence: A perspective of Indian scenario. *Employee Responsibilities and Rights Journal*, 1–16.

12. Sanwal, T., Avasthi, S., & Saxena, S. (2016). E-Commerce and its sway on the minds of young generation. *International Journal of Scientific and Research Publications*, 6(3), 112–117.

12 Cost and Financial Accounting in High-Technology Firms

Lalit Goyal

Assistant Professor, Finance & Accounts, Management Dept. Tula's Institute, Dehradun, NAAC Grade A+ Accredited Institute

Vasim Ahmad

Assistant Professor, Finance & Accounts, Management Dept. Uttaranchal University, Dehradun, NAAC Grade A+ Accredited University

Jugander Kumar

Solution Delivery Lead Analyst, Milwaukee Electric Tool Corporation, 13135 W Lisbon Rd, Brookfield, USA

12.1 INTRODUCTION

Firms are the primary incubators of new technologies. From a firm's point of view, investing in R&D is not an end in itself but rather a means to create outputs (i.e., products and services) that can be offered for sale and turn a profit. This is in addition to pursuing other aims, such as maintaining operations or becoming a technology leader. For the most part, high-tech firms lack the financial resources and other key resources necessary to acquire all of the inputs they need. The inputs needed to efficiently manufacture goods and services will always exceed the resources available; therefore, such businesses must make trade-offs with how they allocate their limited funds. A high-tech company's choices are heavily influenced by financial goals and limits. This paper will reflect on and introduce cost accounting and financial accounting. As a first step towards comprehending the significance of accounting for decisions in high-technology organizations, we will explain a few of the most essential ideas in management and financial accounting. It not only takes into account the private investment investor as a stakeholder, but also zeros in on investors who put money into "cutting edge" or highly innovative fields. This may involve biotechnology or other areas of life science technology, or it may refer to advances in engineering or pharmaceuticals. To do so, the article poses several concerns about the current state of financial statement design [1].

DOI: 10.1201/9781003365525-12

Several major technological developments have been made possible by the data explosion that ushered in the Fourth Industrial Revolution, an era in which cyber-physical systems will alter business. Every firm may capitalize on such significant changes and should pay close attention to how to use them most effectively, but accountants should examine how such technologies can be used strategically to achieve the business goal of the company. Businesses today make use of Big Data, increasing computer power, artificial intelligence, the Internet of Things (IoT), self-driving robots, and blockchain. These technologies have evolved globally in cost and financial accounting [2].

12.1.1 COST AND FINANCIAL ACCOUNTING

The purpose of cost accounting is to collect, analyze, and report on cost data on a regular basis. The primary purpose of this tool is expense estimation and management. Consumers of cost data might benefit from it in a number of ways, including when setting prices, implementing cost-cutting measures, making plans for the future, gauging employee productivity, and a host of other tasks [1].

Financial accounting is indeed the branch of accounting that keeps track of all the monetary transactions of an organization and reports people at the end of the fiscal year in proper formats that increase the readability of financial statements among its users, and cost accounting enhances the efficiency of financial accounting by providing pertinent information that actually resulted in the good decision-making process of the organization. There are a wide variety of audiences for financial data, from upper management to investors [3].

Modern financial management has advanced significantly since the days of traditional corporate finance. As the economy expands and global resources are accessed, the options accessible to finance managers are practically limitless. In recent years, a new age has evolved for chief financial officers in many firms to become finance executives with different titles but comparable roles and responsibilities that are significant in the contemporary context of liberalization, deregulation, and globalization [4].

12.1.2 THE IMPORTANCE OF UNDERSTANDING MONEY IN HIGH-TECH FIRMS GLOBALLY

No matter whether a work title includes management or not, having knowledge of financial matters is vitally important for any businesses or firms globally that supervise operations and provide strategic direction. They could benefit from having an understanding of finances. Understanding finances can be crucial to the success of a company, particularly for those companies whose primary focus is management.

The primary justification for these high-tech businesses to choose to educate themselves in finance is so that they can broaden and deepen their knowledge base. Dual degree holders are becoming increasingly desirable candidates for employment among today's businesses globally. Degrees in engineering and finance are extremely valuable, and both local and international businesses will prioritize candidates who possess these skills [3].

The companies, in India and worldwide, will frequently come up against a significant number of analytical and mathematical challenges. Those who have chosen to pursue a degree in finance after completing their practical training can be confident that they will understand such concerns more quickly than others, which will assist them in resolving a wide range of challenges. It is more likely that employees will be instructed in theoretical and practical challenges, but it is less likely that employees will be instructed in soft skills such as communication, mathematics, and teamwork. However, the acquisition of soft skills is extremely important to one's career, and since they can be learned along the road, this can help individuals differentiate themselves from the competition. In addition to the money gained, gaining a qualification in finance might put a person in a better position financially than people who do not have it [3].

Questions include:

- How can we determine if the production method that we use is efficient if our goal is to develop items or services efficiently; that is, to produce the greatest output with the least amount of input?
- Do any tools exist that can measure how efficient something is?
- How can we boost production of a specific product if there is a rise in the demand for that product?

What additional costs are associated with this, how much additional input is required, and will we still generate a profit even with these changes? After all, it should come as no surprise that the process of conversion differs from company to company, as well as depending on the type the market and the particular technology employed. This is the point where the realms of corporate economics, financial management, and accounting all collide with the technology world. When it comes to finding solutions to technology issues, companies are faced with a wide variety of obstacles and restrictions. It is not enough for a technological solution to satisfy the technological requirements; it must also satisfy the economic and budgetary needs [5].

12.1.3 IMPLEMENTATION OF ACCOUNTING SYSTEMS

The introduction of new management accounting systems is a significant milestone in the history of a young and expanding business. Cross-sectional distinctions in the time it takes for a company to adopt a budget can be explained by a number of factors, including the size of the company, the perceived benefits and costs associated with the budget, and the management style of the top management. Other factors include the adoption of seven different management accounting systems (Figure 12.1).

The figure above explains the indicator's inputs and how the accounting information is consolidated and distributed to the owners, as well as stakeholders such as investors, employees, customers, and suppliers. This is a typical example of financial accounting and reporting. This decision to implement a management planning system is influenced in particular by the availability of venture money, the level of expertise held by the CEO, the existence of a finance adviser, the total

Resource ➡ Indicators ➡ Accounting System ➡ Reports ➡ Use of the Reports

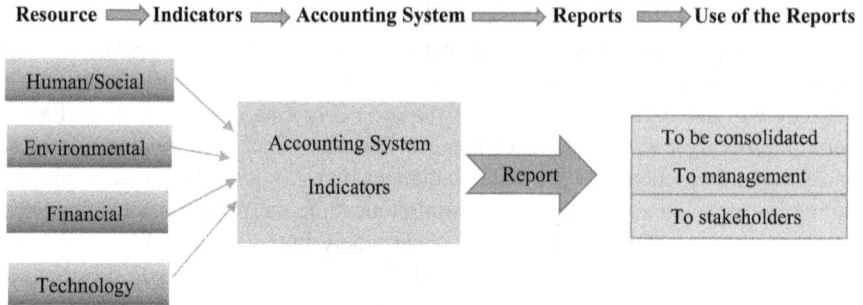

FIGURE 12.1   Accounting system and indicators [6].

number of staff, and the attitudes held by the CEO towards management planning systems. The additional impact of utilizing the services of a finance adviser is an endogenous construct. This extends the amount of time needed to implement operating budgets and impacts the performance of the organization. Also, it provides a draft of a significant rise in the number of employees in the company around the adoption of operating budgets; additionally, the results for additional management accounting systems include cash expenditures, analysis reports, and income statement approval regulations. Capital spending approval policies, profitability analysis, customer profitability, and acquisition costs are provided. At each level of the research, many factors have an impact, including industry biotechnology, information systems, and non-tech factors [7].

There are many different kinds of businesses, but in terms of financial accounting, there are fundamentally no differences between them. Companies in the technology industry are required to adhere to the same accounting requirements as other businesses. Additionally, the fundamentals of management control and management accounting are universally the same across all businesses. There is no difference between high-tech businesses and other types of businesses. There are, however, some distinctions to take into account, such as the unpredictability of customer demand and the limitations imposed by finances. This is especially the case when the high-tech business in question is a start-up that does not have a solid financial and commercial reputation, that has a limited amount of financial resources, and that has large expenditures for the creation of new products and services [5].

12.2   ORGANIZATIONS IN FINANCIAL ACCOUNTING AND INFORMATION TECHNOLOGY

The principal source of information that can be independently validated and made available to financiers is provided by financial accounting. This information is about the manager's performance. Therefore, it is abundantly evident that financial accounting and corporate governance are inextricably tied to one another. In point of fact, if one does not approach financial accounting from the perspective of corporate governance, it can be challenging to comprehend a number of the most important aspects of the discipline, including the application of historical costs, this same reliability criterion, the moment of realization principle, and the conservatism principle. If there were no issues with governance, the function of financial accounting would be simplified to supplying investors with the return and risk information that is necessary to make the decision on the optimal allocation [7].

Accounting's end goal can be broken down into two sub-goals: control and decision making. Decision making to manage mental accounting entails both strategic and operational choices, while control encompasses planning, administrative, and cultural controls, as well as compensation systems. In financial accounting, the terms "control" and "decision making" refer, respectively, to stewardship accounting (wherein management is responsible to stakeholders, particularly to investors, for the resources they are entrusted with) and the "valuation focus," wherein investors are given information to use in making educated investment decisions. The decision-making process and supervision may seem like quite distinct ends to achieve, but the institutions and practices of management accounting and financial accounting share a common goal [8].

12.2.1 INTEGRATION OF FINANCIAL AND MANAGEMENT ACCOUNTING

The original use of managing mental accounting was for controlling the past, but it has since shifted to include the use of forward-looking information systems for strategic planning, control, and decision making. Due to managers' need for retrospective data in order to comprehend performance and regulate accountability, manage mental accounting has traditionally concentrated on yearly controls in stable and constrained competitive operating contexts [8]. While the managed mental accounting information has become more forward-looking as a result of recent accounting changes, certain parts, such as budgeting and public investment projections, have always been included. Corporate today is defined by a number of ongoing trends, including globalized competitiveness, business networks, and the growing significance of securities as mechanisms for allocating financial resources. This has led to a rise in the demand for future-focused, forward-looking data to aid in management accounting strategy planning and decision making [9].

12.2.2 THE VALUATION ROLE OF ACCOUNTING INFORMATION IN FIRMS

Investors in the company will be interested in learning how the money will be spent, and creditors will be concerned with whether or not they will be repaid for their loans. Because of this, the financial stability of the company is a big concern for possible buyers, investors in venture funding or private equity, private corporations wishing to invest, and other potential investors. Therefore, whether these involve raising finance, borrowing the money, or deciding to sell the company in order to prove its worth as a reduced business asset, it is important to consider all of these options [9].

The economic value of a company can be estimated through the process of business valuation. This is accomplished by a procedure that assesses the value of the firm by examining all aspects of the company, most notably its finances and stability. No matter what method of business valuation the value or analyst uses—market capitalization, the occasions revenue pricing method, this same earnings multiplier method, the discounted cash flows method, book value, valuation of assets, or liquidation value—they will be required to analyze the financial statements [8].

The company's market value will not be reflected in its financial statements, which include tax returns, statements of income, statements of cash flows, accounting records, assets and liabilities valuation reports, statements of income, and declarations of stockholders' equity. On the other hand, given they are reporting on the transactions that the organization has already documented in the past, they will be utilized as a foundation for transactions that are perceived to take place in the future. An increase in a company's valuation will occur when it can demonstrate it is profitable and will generate positive cash flows in the future [10].

Therefore, accounting and bookkeeping practices that are not up to par can put value at risk, which is especially true right before a transaction is set to take place. And buyers and investors will still want to look just at financials and feel sure about their investment even if they won't be the ones performing the evaluation on the firm themselves. This is because they will want to feel more confident about the investment [11].

A well-controlled budget and secure financial position are the direct results of accurate accounting. The company will be exposed as having poor financial management and judgment if it is unable to provide accurate financial statements, such as a report of debits and credits, a net income tax statement, a cash flow statement, accounting records that show payables and accounts receivable, and financial information. Given that these are indications of a ship that is going down, investors will most likely pull out before the experts have finished valuing the company. After all, if they aren't keeping an eye on the money and enforcing accounting rules, it's highly doubtful that they will be able to keep costs under control and take proactive steps to improve the bottom line [12].

When the finance and accounting function of an organization is inefficient, it leaves the company open to the possibility of making mistakes and causing delays, both of which can lead to additional labour, overpayments, and a reduction in income. The organization is at significant risk of becoming the victim of fraud as a result of outdated systems, as well as a lack of internal controls. An inefficient and inept finance department is an indication of poor business management, which in turn will have an influence on the value of the company. All of these warning indicators will ultimately have an impact. When a company is poorly managed and does not have access to business information, it is doubtful that the investment will be profitable and worthwhile. On the other side, financial reporting that is both accurate and efficient is a sign that the company is well positioned for expansion [11].

12.3   COST MANAGEMENT IN FIRMS

The practice of monitoring and controlling a company's monetary resources while it is being operated is referred to as cost management. Globalization is producing a world in which individual nation-states are interdependent and interrelated to a greater degree. In the previous couple of decades, significant and dramatic developments in the business environment have contributed to a high degree of complexity, instability, and unpredictability in the environment in which modern enterprises carry out their economic purposes. The trends towards globalization and the elimination of national obstacles will undoubtedly result in rapid intensification of international competitiveness [13]. When efficient tactics for managing expenses are put into action, it ensures that a company has effective cost management measures, which in turn helps the company to have a suitable budget with which to deal with the many activities that the firm engages [12]. It is possible to adopt cost management techniques either on a project-by-project basis or for the entire firm. The expenses of an organization may be kept under control, and the organization's revenues can be maximized to the greatest extent feasible if the costs are appropriately controlled. When expenses are tracked at each stage of the process, both the planning and allocation of resources will be more precise [14].

The Key Elements

- The administration and control of a company's costs are referred to as cost management. This is done so that companies can spend less money and make more, hence increasing their profits.
- The process of management has as its goal not the reduction of costs, but rather the limitation of cost cutting to a level that does not compromise product quality while still achieving the desired level of savings [14].
- Cost management is an essential aspect of running a successful organization since it involves elements such as planning, communication, motivation, evaluation, and decision making.
- The primary responsibilities of the process of cost management are broken down as follows: resource allocation, cost estimation, cost budgeting, and cost control.

The term "cost management" refers to the process of controlling the expenses that are connected with running a firm. For instance, from the point of creating items to the point of delivering them to customers, a company is required to spend money on raw resources and shipment methods. All of these expenditures contribute to the overall expenses that businesses must bear in order to generate revenue from the sale of finished items [12].

Businesses often cut corners on the quality of the raw materials used in the production of their goods in an effort to lower their overall operating expenses. However, they need to be aware that if they cut corners just on quality of the items they sell, it will only result in losses for the organizations they work for, whether those losses come in the form of decreased sales or a damaged reputation [15].

12.3.1   COST MANAGEMENT PROCESS

The manufacturers' cost management strategy has evolved from "making a living by selling what you create," based on a model of creating items in Japan and exporting them elsewhere, to "making a living by constructing what you can sell," based on a globally optimized production model. In this climate, it is essential to have product plans for designing items that are in demand on international markets with varying levels of economic power, culture, and infrastructure, and then producing and marketing these products at competitive prices. Hitachi provides cost management based on product profit planning as a solution. Against the backdrop of a dynamic business climate, each company must prepare by utilizing tactics that take into account its individual product qualities, manufacturing situations, and market competition. Cost management is an efficient method for overcoming several obstacles [16].

There are processes and functions that exist, regardless of whether one is attempting to manage the expenses of a specific project or the costs of the business as a whole, to ensure that both goals are met. They can be divided up into the following four categories:

The initial stage is to divide up the available resources. In order to complete the process, we will need to identify and allocate the resources and schedule them. The next thing that the experts do is make a list of the resources that are necessary at each stage of the development of the project or product. In addition to this, they are responsible for the costs that are involved with the acquisition and provision of those resources [15].

The following is the estimation of the costs. It involves making forecasts on the amount, cost, and pricing of the resources that will be needed for the project. In this process, corporations use a variety of methods to transform the information they have about the project that is not financial into information that is financial. These outputs are converted into inputs for the primary purposes of cost planning and analysis for the overall project [17].

The following step is indeed cost budgeting, which is an integral component of the earlier cost estimation step. During the process, the financial needs of a project are determined, as well as the overall cost is assigned to the appropriate cost accounts. After that, the relevant costs turn into a foundation for cost control, which can be used later on to make comparisons. The planning accounts for every possible expense, from the expenses of materials and labour to the expenditures of administration and software.

The comparison and appraisal of expense accounts come in last, but it's still an important part of the cost control process. During this stage of the process, the companies investigate the disparity between the anticipated cost and the actual cost. The variations enable the organization to secure the lowest feasible cost while maintaining the highest possible standard for the final product (Figure 12.2) [17].

As explained, the cost management functions in the above figure, the corporations keep track of the expenditures and performances of the project and evaluate them in relation to its advancement; they then adjust their strategies accordingly [15]. The following is the estimation of the costs. It involves making forecasts on the amount, cost, and pricing of the resources that will be needed for the project. In this process, corporations use a variety of methods to transform the information they have about the project that is not financial into information that is financial. These outputs are converted into inputs for the primary purposes of cost planning and analysis for the overall project [17].

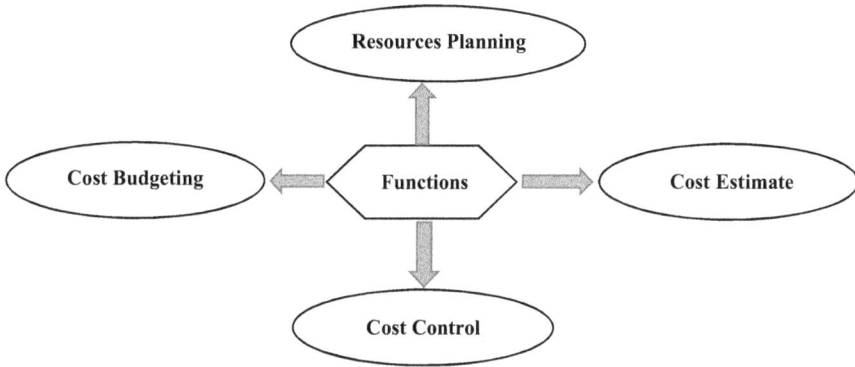

FIGURE 12.2   Cost management process.

12.3.2   COSTING AND INTANGIBLE ASSETS

Intangible assets, also known simply as intangibles, are assets that do not exist in a tangible form but nonetheless provide value to the owner. There are two primary classifications of intangibles, namely, value improvement and identified intangibles [18].

Intellectual property (IP), including trademarks, customer connections, and contracts, are examples of things that fall under the category of identifiable intangibles. These assets often have legal property rights and can be sold or severed from the business if the owner so chooses. The owners have several options for making use of these assets within the company, including charging license fees or royalty, or selling these assets. In the other category are the intangibles, which include things like work processes, skilled leadership, and trained personnel. These things create competitive advantages and boost the value of a going concern, but they cannot be sold independently from the firm.

Despite the growing importance of intangible assets to the value of a company, it can be challenging to accurately reflect these assets in financial statements due to the way accounting is currently practiced. This knowledge gap has the potential to have a negative impact on valuations [18].

12.3.3   HOW IT HELPS THE FIRMS

Intangible assets are things that are listed on a company's balance sheet but do not exist in physical form. These assets may have monetary or business value buried inside them, but the physical form is not existent. Companies gain a competitive advantage thanks to the intangible assets' ability to undertake operations in a manner that is distinct from that of their rivals. Intangible assets can take many forms, including patents, trademarks, and copyrights, all of which fall under the category of intellectual property. There are two ways for a company to obtain access to intangibles: either by generating intangibles itself or by purchasing intangibles from other companies [19].

It is also possible for the intangible worth of a business to be concealed within the value of the brand that the company possesses. There are numerous unique

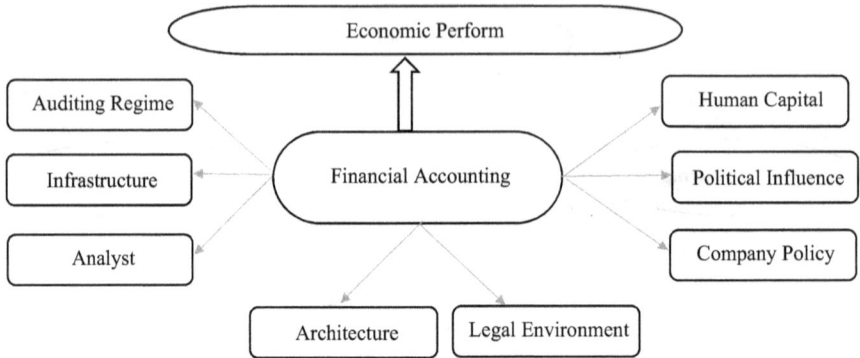

FIGURE 12.3 Global brand value of high technology firms [20].

selling points (USPs) that a company may possess, each of which may be regarded as contributing to the company's overall intangible value. It is noteworthy to notice that five of the world's top ten brands are associated with the technology industry. This further demonstrates the extent to which the technological sector has permeated our day-to-day lives and the implications. The most valuable brands in the world are listed below (Figure 12.3).

It's noteworthy to observe that the technology sector accounts for 5 of the top 10 global brands. This further demonstrates the impact that technology is having on our daily life. The recent rise in the number of initial public offerings (IPOs) is more evidence that intangible assets are more important than they have ever been in this day and age [21].

The percentage of intangible assets held by S&P 500 businesses has been growing steadily over the past half century, as the pie chart below demonstrates. It is common knowledge that leading technology businesses such as Microsoft and Apple may brag about having the largest worth of intangible assets in the entire globe. Their intangible assets consist of customer-related assets, assets connected to technology and marketing, assets related to contracts, and assets related to customers [22].

12.3.4 FINANCIAL ACCOUNTING

Managers and investors benefit from having access to the financial accounting data of enterprises and their competitors so that they can identify and evaluate investment opportunities. When an economy does not have access to information that is both reliable and easily accessible, it becomes more difficult for human and financial resources to flow into sectors that are anticipated to have high returns or away from areas that have bad prospects [21].

Even in situations where there are no agency conflicts between and investors, the availability of high-quality financial accounting data can improve efficiency by making it easier for managers and investors to recognize opportunities to create value with less room for error. This directly leads to a more precise allocation of cash to the purposes with the highest value, as the indication shows. A reduced risk of estimation can also result in a lower cost of capital, which is another factor that contributes to economic performance [23].

The information that financial accounting systems provide regarding prospects for investment is transparent and straightforward. Just on the basis of profit margins stated by other companies, for instance, managers or companies who are considering entering the market can locate potentially lucrative new investment prospects, candidates for acquisition, or strategic developments. The informational function performed by a company's stock price is another area that is supported by financial accounting systems. According to the argument put up, the mere existence of active securities markets depends on the maintenance of a robust financial accounting system that places a significant emphasis on credibility and responsibility [24].

12.3.5 FINANCIAL ACCOUNTING DATA INFLUENCES ECONOMIC OUTCOMES

One way to look at a corporation as a nexus of agreements with the goal of reducing the expenses associated with contracting. Parties interested in entering into contracts with the company want information not only about the company's capacity to fulfill the terms of contracts, but also about the company's overall compliance, including its contractual commitments. The information that is provided by financial accounting provides a crucial mathematical representation of particular firms, which is used to support a diverse array of contractual connections. The information obtained through financial accounting also improves the information environment in a more broad sense since it disciplines the unaudited disclosures made by management and provides input into the data processing activities carried out by other parties [25].

In addition to having an effect on the cost of capital at which cash flows are discounted, the quality of financial disclosure can have an immediate and direct impact on the cash flows of companies. As shown in the exhibit, our theory postulates that there are three different ways in which information on financial accounting can boost economic performance. To begin, managers and investors are helped in the process of finding and analyzing potential investment possibilities by the financial accounting information of enterprises and their competitors. When an economy does not have access to information that is both reliable and easily accessible, it becomes more difficult for human and financial resources to flow into sectors that are anticipated to have good yields and away from areas that have bad prospects [23].

Even in situations where there are no agent conflicts between managers and investors, the availability of high-quality financial accounting data can improve efficiency by making it easier for managers and investors to recognize opportunities to create value with less room for error. This directly leads to a more accurate allocation of money to the uses that have the highest value. Lower estimation risk can also lower the cost of capital, which further contributes to the performance of the economy.

The information that financial accounting systems provide regarding prospects for investment is transparent and straightforward. As mentioned previously, just on the basis of the profit margins stated by other companies, for instance, managers or companies who are considering entering the market can locate potentially lucrative new investment prospects, candidates for acquisition, or strategic developments. The informational function performed by a company's stock price is another area that is supported by financial accounting systems. In order for there to even be active securities markets, there must first be in place a rigorous system of financial accounting that places a heavy emphasis on both credibility and accountability [22].

The corporate climate has a considerable impact on accounting and financial reporting policies worldwide, particularly in Egypt. Therefore, it is vital to investigate the tradition of conservatism in accounting and its influence on corporate performance indicators in Egyptian businesses. Continuous independent variables include leverage, business efficiency, expenditure decisions, accounting conservatism, firm size, and cash management. The impact of accounting conservatism on financial performance metrics in Jordanian insurance firms (Figure 12.4) [26].

The figure above represents how financial accounting affects various factors and leads to economic development as well. Financial accounting plays various roles in uplifting the financial economic performs. The Stock Markets that are efficient, as measured by the extent to which stock prices represent all available public information and also aggregate this same private information held by individual investors, should presumably communicate this aggregate information to managers as well as existing and potential investors [24].

This explicitly models a role for stock prices in which strategy plays a directing function. In these models, the stock price impounds private, choice-relevant information that is not already known by managers, managers' investment decisions respond to the appearance of this new information in price, as well as the market correctly anticipates managers' decision strategies when setting price [15].

The governance role that financial accounting information plays is the second route through which we anticipate financial accounting information to improve economic performance. The transactions of a company need to be recorded in order to create the financial statements for that company. The financial statements are derived from these bookkeeping records, which offer the fundamental statistics used in their construction. It is impossible for financial statements to conform to the criteria established by US GAAP or IFRS if transactions are not properly and accurately recorded. This suggests that the recording of transactions and the

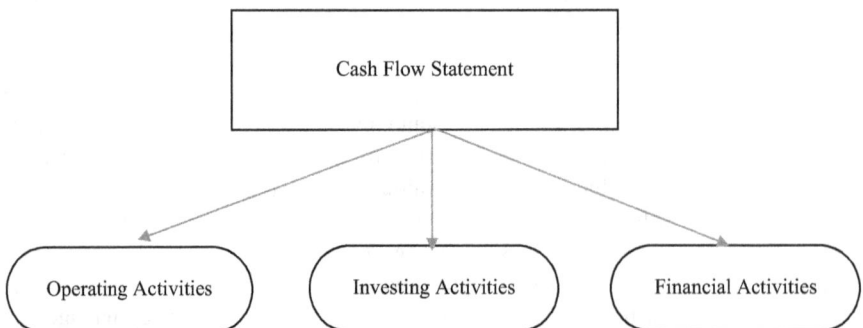

FIGURE 12.4 Financial accounting system.

assigning of indirect costs to finished goods or services must conform to the accounting standards in question [27].

The International Financial Reporting Standards (IFRS) have already been accepted as the benchmark for financial and accounting reporting on a global scale, and their implementation in the region is becoming increasingly common. These standards make certain that representatives from a diverse range of domains, businesses, and nations have access to a standardized language for reporting financial and accounting information. Participants will walk away from the programme with a deeper comprehension of the performance-related and advanced accounting, reporting, and auditing difficulties presented by the International Financial Reporting Standards. The accounting technique and the final production of financial statements both make use of the accounting standards that have been developed and articulated [28].

Individual discussions of each of these three financial statements will provide an introduction to the fundamentals of financial accounting. The connection that exists between the three components will next be discussed, and it will be demonstrated that the three financial statements the balance sheet, the income statement, and the cash flow statement provide the information when taken as a whole. The governance function of financial accounting information makes a direct contribution to economic performance through the disciplined management of existing assets (for example, the timely abandonment of losing projects), improved project selection, and reduced expropriation of investors' wealth by managers. We also take into consideration the likelihood that the disclosure of financial accounting information reduces the risk premium that investors want as compensation for the chance of suffering a loss as a result of opportunistic managers seizing opportunities [27].

On the other hand, we want to warn that the effect of greater governance on the return rate that investors want is a very delicate one. It argues that the influence increased governance has on the make purchase decisions based on return on equity varies on the type of change that was made. For instance, improved governance can be seen as a reduction in the private benefits that supervisors can extract from the company or as a reduction in the legal and auditing expenses that shareholders should bear to prevent managerial opportunism. These are just two examples of how improved governance can be manifested in practice. It is possible for these two shifts to have opposing effects on the equilibrium stock returns that have been seen, and the magnitude of these effects is dependent on the degree to which international equities markets are segmented [29].

12.3.6   CASH FLOW STATEMENT

A cash flow statement is a type of financial statement that gives aggregate information regarding all cash inflows a firm receives from its ongoing activities and external investment sources. Cash inflows can come from either internal or external sources. In addition to this, it takes into account any and all cash outflows incurred within the specified time period to cover the costs of business activities and investments [30].

The financial statements of a company provide investors and analysts with a picture of all the transactions that take place within the firm, where each transaction contributes in some way to the company's overall success. It is generally agreed that the cash flow statement is the most straightforward of all the financial statements. This is because it tracks the cash that is created by the company in three primary ways: through its activities, investments, and financing. The total of these three components is what is known as the net cash flow [31].

In addition, the cash flow statement is included in the financial statements. In some textbooks, the statements of sources and uses are handled independently from the statement of cash flow. During the course of an accounting period, both accounts provide an account of how a company acquires its cash (referred to as "sources") and how it uses its cash (referred to as "uses"). Only the statement of cash flows will be covered in this section [32]. This statement provides data on the connection between the change in cash holdings and the net income that was earned. The cash flow statement is used to report on previous cash flows as assistance in evaluating the generation and use of cash, determining the ability of a company to pay dividends and interest and to repay loans when they are due, or in anticipating future cash flows and operating outcomes (Figure 12.5).

Cash flow statement components are explained in the above figure. It's possible at this time to be wondering if there is a need for a cash flow statement in addition to the income statement. The fact that the income statement includes accruals like depreciation that do not generate cash flows is one of the most significant issues with the document. In this calculation, those cash flows that are directly related to the purchase of goods are taken into account [33]. Other cash flows, including such investments or financial transactions, are not included in the income statement. Despite the fact that these cash flows do have an impact on the company's cash position, these cash flows are not included. Even in the best-case scenario, the income statement only gives a partial picture of the cash flows. On the other hand, the cash flow statement details all cash flows. This includes not just cash flows resulting from operating operations, which are the major concern of the financial statements, but also cash flows resulting through investments and financing activities [33].

12.3.7 Accounting Helps in High-tech Firms

As per the analysis, here are a few barriers that have been found in the development of high-tech firms for adapting the accounting. Actually, the major obstacle to the growth of high-tech enterprises is a lack of personal capital. Unexpectedly, the shortage of highly qualified staff for high-tech enterprises moved to the top spot in the ranking of development hurdles, overtaking the lack of financial resources. The second and third-ranked studies at the Higher School of Economics are hampered by a lack of government funding and the high cost of inventions (Table 12.1).

After closely analyzing the functions of the cost and finance accounting system in the company, we can create a graph of the key elements of the accounting as well. Now we will discuss how accounting helps in this firm and why it's so important to keep an account of the cost and finance of all data [31].

After employment in accounting was outsourced to developing countries, the profession is now preparing for new threats from developed nations. In order to rise to the occasion, the accounting industry in India requires a discernible improvement in both output and quality. Indian accountants need to be knowledgeable about the

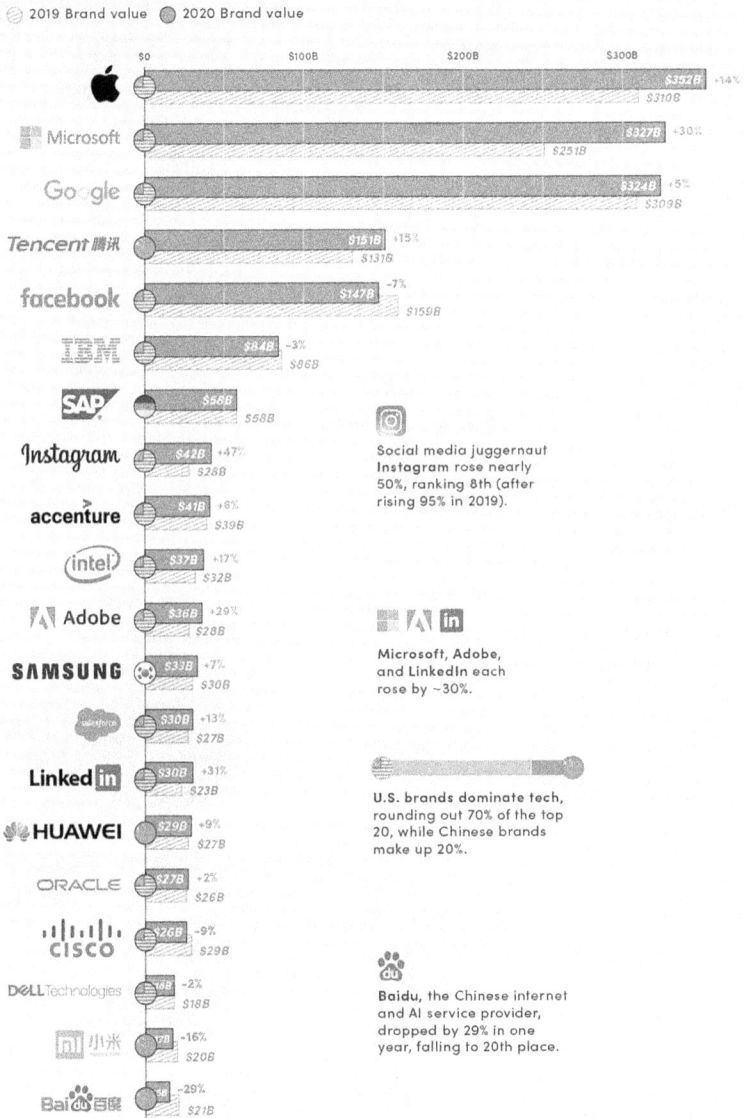

2019 Brand value   2020 Brand value

$0 $100B $200B $300B

Apple $352B +14% / $310B

Microsoft $327B +30% / $251B

Google $324B +5% / $309B

Tencent 腾讯 $151B +15% / $131B

facebook $147B -7% / $159B

IBM $84B -3% / $86B

SAP $56B / $58B

Instagram $42B +47% / $28B

Social media juggernaut Instagram rose nearly 50%, ranking 8th (after rising 95% in 2019).

accenture $41B +6% / $39B

intel $37B +17% / $32B

Adobe $36B +29% / $28B

SAMSUNG $33B +7% / $30B

salesforce $30B +13% / $27B

LinkedIn $30B +31% / $23B

Microsoft, Adobe, and LinkedIn each rose by ~30%.

HUAWEI $29B +9% / $27B

ORACLE $27B +2% / $26B

CISCO $26B -9% / $29B

U.S. brands dominate tech, rounding out 70% of the top 20, while Chinese brands make up 20%.

DELL Technologies $18B -2% / $18B

小米 $17B -16% / $20B

Baidu 百度 $15B -29% / $21B

Baidu, the Chinese internet and AI service provider, dropped by 29% in one year, falling to 20th place.

FIGURE 12.5 Cash flow statement.

most recent developments in the accounting industry to be able to meet these needs. This includes knowledge of new techniques for voucher entries, new methods for accounting of departments and branches, and new systems for the automation of accounting work. This should only be achievable with accountant interfaces brought about by 21st-century information technology (IT), which is playing a key role in the period of globalization. The significance of IT refers to a broad category that

TABLE 12.1
Barriers of Development

Barriers for the Development	Statistics
Lack of own funds	1
Lack of qualified personnel	4
Imperfection of legislation	–
Low demand for new products	9–11
High economic risk	5
Lack of information on sales markets	6–7
Long payback periods	6–7
High cost of innovation	2
Underdevelopment of the technology market	8
Underdevelopment of innovation infrastructure	9–11
Lack of information about new technologies	–
Lack of financial support from the state	3
Weak cooperative ties	9–11

includes recent advances in computing and communication technology. IT workers are responsible for the design, development, and management of these systems, which rely heavily on computer hardware, software, and the Internet [32].

12.3.8 THE KEY FEATURE OF ACCOUNTING

It enhances the ability of research and extension professionals in accounting to organize, store, retrieve, and exchange accounting information. It helps to develop some kind of system for the exchange of information. It creates a database for accounting purposes that is easy to use and allows for data-based decision making [32].

It is not difficult to gain access to the information warehouse. The accounting data will be readily available to everyone. With the right application of information technology in accounting, a single accountant may handle the needs of a number of different businesses.

12.3.9 THE IMPORTANCE OF ACCOUNTING INTERPRETATION

The practice of reviewing accounts in great detail in order to discover the kind of business activities that have been recorded in them for a specific accounting period is referred to as account interpretation. The management team needs the information received through account interpretation in order to properly plan and control the activities of the firm. Interpretation is an essential tool for management in the field of accounting since it reveals trends as well as anomalies that are odd or unexpected [33].

As a result of the ongoing development of financial transactions, new scenarios are emerging that might not have been anticipated by the accounting standards that are now in place. Accounting boards, like the Financial Accounting Standards

Board (FASB), the American Institute of Certified Public Accountants (AICPA), as well as the International Accounting Standards Board (IASB), have the option of issuing an interpretation that outlines the best practices for accounting whenever questions arise. The International Financial Reporting Standards Foundation (IFRSF) acts as the governing body for the International Accounting Standards Board (IASB) [34].

It is possible that an entirely new standard will be issued for a category of financial transactions that were: not previously in existence. This could happen, for instance, for industries that have only recently come into existence, such as certain subsets of the technology sector in the last two decades. An accounting interpretation provides clarity, which increases the likelihood that the reported financial data are meaningful, accurate, and comparable among different firms. It is in the best interest of investors to ensure that everyone adheres to the same guidelines, as this enables them to make more educated choices on the stocks on which they should spend their money [34].

12.4 CONCLUSION

The practice of accounting in high-tech companies (worldwide) requires making crucial decisions concerning monetary matters, such as evaluating capital investments, allocating budget funds, and monitoring corporate performance. It gives the company's stakeholders information about the company's financial situation. Accounting is the process of measuring and summarizing the operations of a company and communicating those measurements and summaries to management as well as any other parties who may be interested. In the day-to-day operations of a high-tech company, the decision-making process is heavily influenced by financial objectives and limitations. Both managerial accounting and financial accounting are covered in this article in an introductory fashion.

The data are collected through the accounting process and presented in a variety of different reports. Accountants provide assistance in interpreting the meaning of the reports and offer methods to use the details to find solutions to difficulties. As a first step towards comprehending the significance of accounting for the managerial decision-making process in high-technology companies, we will begin by presenting a few of the management and financial accounting concepts that we believe to be the most significant. This is due to the expansive nature of accounting. When developing a sales plan and a predicted product mix, it is absolutely necessary to have an accurate accounting of the manufacturing expenses associated with each product. It is quite likely that each commodity will have a distinct contribution to the overall gross profit, and management will need to set sales goals for each individual item in order to achieve the level of overall gross profit required to pay for operating expenses and produce the desired net profit. It has been determined that financial statements, both managerial and financial, are necessary components of effective business management in any firm or organization. There is no alternative to this. A recipe for disaster is the absence of information that is both accurate and up to date regarding the efficiency with which a firm is being handled.

REFERENCES

1. Anilowski, C., Feng, M., & Skinner, D. J. Does earnings guidance affect market returns? The nature and information content of aggregate earnings guidance. *Journal of Accounting and Economics* (2007)

2. Marr, B. (2020, July 27). The 6 biggest technology trends in accounting and finance. Forbes. https://www.forbes.com/sites/bernardmarr/2020/07/27/the-6-biggest-technology-trends-in-accounting-and-finance/?sh=3bb2127d3c7e

3. Arnold, V., Benford, T., Canada, J., & Sutton, S. G. The role of strategic enterprise risk management and organizational flexibility in easing new regulatory compliance. *International Journal of Accounting Information Systems* (2011)

4. (N.d.-m). Icmai. In. Retrieved December 29, 2022, from https://icmai.in/upload/Students/Syllabus2016/Inter/Paper-10-Oct.pdf

5. Cadez, S., & Guilding, C. An exploratory investigation of an integrated contingency model of strategic management accounting. *Accounting, Organizations and Society* (2008)

6. Fagerström, A., Hartwig, F., & Cunningham, G. (2017). Accounting and Auditing of Sustainability: Sustainable Indicator Accounting (SIA). *Sustainability (New Rochelle, N.Y.)*, *10*(1), 45–52. doi:10.1089/sus.2017.29080.af

7. Deakins, D., & Hussain, G. Financial information, the banker and the small business: a comment. (1994)

8. Dushnitsky, G., & Lenox, M. J. When do incumbents learn from entrepreneurial ventures? Corporate venture capital and investing firm innovation rates. (2005)

9. Bebchuk, L., & Fried, J. *Pay without performance: the unfulfilled promise of executive compensation.* Cambridge: Harvard University Press; (2004)

10. Bhimani, A., & Keshtvarz, M. H. British management accountants: strategically oriented. *Journal of Cost Management* (1999)

11. Press, G., & Weintrop, J. "Accounting-Based Constraints in Public and Private Debt Arrangements: Their Association with Leverage and Impact on Accounting Choice." *Journal of Accounting and Economics* 12, (1990)

12. Kang, J., & Shivdisani, A. "Firm Performance, Corporate Governance, and Top-Executive Turnover in Japan." *Journal of Financial Economics* 38, (1995)

13. Stefanović, R. J. (n.d.). Cost accounting and company management in a world without walls. Culturaldiplomacy.org. Retrieved December 29, 2022, from https://www.culturaldiplomacy.org/academy/content/pdf/participant-papers/2010www/Cost_Accounting_and_Company_Management_in_a_World_without_Walls.pdf

14. Brazel, J. F., & Dang, L. The effect of ERP System implementations on the management of earnings and earnings release dates. *J Inf Syst* (2008)

15. Burns, J., & Vaivio, J. Management accounting changes. *Manage Account Res* (2001)

16. Fukute, K., Ashikawa, M., Fukuyama, S., & Yamamura, K. (n.d.). Global cost management solution supporting business strategy. *Hitachi.com*. Retrieved December 29, 2022, from https://www.hitachi.com/rev/pdf/2015/r2015_05_110.pdf

17. Eccles, R. G., Herz, R. H., Keegan, E. M., & Phillips, D. M. H. *The value reporting revolution: moving beyond the earnings game.* Chichester: Wiley. (2001)

18. Parker, L. D. Interpreting interpretive accounting research. *Critical Perspectives on Accounting*, (2008)

19. Graham, B., & Meredith, S. B. *The interpretation of financial statements.* London: Harper & Brothers. (1937)

20. Wood, T. (2020, August 4). Ranked: The world's 20 biggest tech giants, by brand value. Retrieved March 26, 2023, from Visual Capitalist website: https://www.visualcapitalist.com/the-worlds-tech-giants-ranked/

21. Hand, J. R. M. The value relevance of financial statements in the venture capital market. *Accounting Review*, (2005)

22. Great Britain Parliament. *Companies Act*. London: HMSO (2006).
23. Lambert, Richard A., Contracting Theory and Accounting. JAE Rochester Conference April 2000. (2001)
24. Rajan, R., & Zingales, L. "Financial Dependence and Growth." *American Economic Review* 88, (1998)
25. Oliveira, L., Rodrigues, L. L., & Craig, R. Intangible assets and value relevance: evidence from the Portuguese stock exchange. *The British Accounting* (2010)
26. Sana'a, N. M. The effect of accounting conservatism on Financial Performance Indicators in the Jordanian Insurance Companies. *Journal of Internet Banking and Commerce*. 2016; 21(1):1–16
27. Jensen, M., & Murphy, K. "Performance Pay and Top Management Incentives." *Journal of Political Economy*. (1990)
28. International Journal of Management Excellence (Vol. 8, Issue 2). (2017). *Techmind Research Society*.
29. Thomas, G., & James, D. Reinventing grounded theory: some questions about theory, ground and discovery. *British Educational Research Journal*, (2006)
30. Ueda, M., & Hirukawa, M.. Venture capital and industrial 'innovation'. *SSRN eLibrary*. (2008)
31. Schertler, A. Knowledge capital and venture capital investments: new evidence from European panel data. *German Economic Review*. (2007)
32. Ryan, B., Scapens, R. W., & Theobald, M.. Research method and methodology in finance and accounting. *South Western Publications*. (2002)
33. Wilkins, J., Van Wegen, B., & De Hoog, R. Understanding and valuing knowledge assets: Overview and method. *Expert Systems With Applications*. (1997)
34. Scapens, R. W. Doing Case Study Research. In C. Humphrey & B. Lee (Eds.), *The real life guide to accounting research: A behind-the-scenes view of using qualitative research methods*. Oxford: Elsevier. (2004)

13 Kaizen

A Philosophy for Survival and Revival of SMEs after COVID-19

Meenakshi Tyagi
Associate Professor, KIET Group of Institutions, Delhi- NCR, Ghaziabad

Sapna Yadav
Jaypee Business School, JIIT, Uttar Pradesh

Shivani Agarwal
Assistant Professor, KIET School of Management at KIET Group of Institutions, Ghaziabad

Himani Grewal
Assistant Professor, Moradabad Institute of Technology, Moradabad

Ruchita Gautam
Professor Department of Electronics and Communication Engineering KIET Group of Institutions, Ghaziabad

13.1 INTRODUCTION

Though things have started becoming normal now, during COVID-19 new things emerged, and umpteen number of changes came into existence. These changes were positive as well as negative as the IT sector remained quite blessed with higher profits. On the other side, manufacturing units had to face a number of problems in terms of cost, production, human resource availability, on-time delivery challenges, and many more. Due to all these problems, small and medium enterprises (SMEs) in the manufacturing sector were not able to cover the cost of production, and this shook these units badly. In this situation, the first thing that comes to mind for decision makers is cost cutting. As per the traditional approach, companies would think that the only way to increase profits is to cut costs, so they would start cost cutting internally and externally. But they forget that there is a difference between cost and wastage.

DOI: 10.1201/9781003365525-13

As per the modern approach, companies need to identify the wastes and eliminate these wastes from the system instead of cutting costs. Cost is an essential expense to produce a particular product or service. Reduction in salaries, facilities, product quality, packaging, and other services is not a way to increase profits. Rather, this will be the cause of deterioration in goodwill of the company and market share. If companies do this without any systematic assessment, it will result in the loss of various performance indicators, eventually demotivating the entire team and customers, and also shutting down the company in severe conditions. Without proper assessment, cost cutting can be like cutting muscles along with unwanted fat, which results in serious illness.

Tough times are great educator because they teach companies to seek new ways and work with new paradigms. The real heroes only emerge during this time. Companies should understand and learn ways to identify the components of cost and waste and then disregard the waste so that the product can be delivered with more value.

Cost is the outcome of an activity, and it has been observed that any activity has two components. One is called *value adding*, and the second is called *non-value adding*. Activities that do not add any value to the product or service are known as non-value adding activities. They eat effort and time; thus, a company must eradicate these activities from the system and focus on value adding activities. To better understand, let's have a look at the below equations, while deciding the selling price of a product:

1. Selling price of product = Cost (all types – fixed and variable) + Profit
2. Selling price of product = All activities involved + Profit
3. Selling price of product = Value adding activities + Non-value adding activities + Profit

So, companies must analyze all the activities carefully to categorize value added and non-value added activities and then remove non-value-adding activities that are the cause of waste.

Study and research proved that most of the processes (Process – set of activities to transform input to output) are only 15 percent value adding, and 85 percent of activities are non-value adding activities, which are the cause of waste.

To analyze the process of what are value adding and non-value adding activities, Kaizen methodology can be implemented. This methodology will help to identify and management of non-value adding activities. Kaizen is a Japanese word consisting of two words, Kai + Zen, the literal meaning of Kai in English is change, and Zen is betterment, so put together is change for betterment, which brings continuous improvement. Improvement is not possible without change, so to make continuous improvement in a system, Kaizen philosophy must be implemented to bring cultural changes in organizations as well.

13.2   REVIEW OF LITERATURE

Carvalho and Reis (2021) showed that the benefits of adopting Kaizen in a production system are productivity improvement and operational waste reduction, with a subsequent increase in effectiveness.

Tyagi et al. (2020) wrote about how Kaizen is helpful to differentiate value-adding and wasteful activities. Georgise and Mindaye (2020) emphasized adopting the culture of

Kaizen and developing the habit of continual improvement in the system by using Kaizen to improve productivity and devise strategies. Sehleanu and Flore (2019) advocated for the implementation of Kaizen to diagnose and cut off waste and improve the production process. Manuel F. Suárez Barraza et al. highlighted the benefits of Kaizen implementation in SMEs in Latin America. Iwao (2017) threw light on Kaizen philosophy and stated that Kaizen is a holistic process of obtaining benefits at all levels in an organization. Manjunath Shettar et al. emphasized the relevance of the Kaizen approach and how Kaizen works to transform high-cost companies into high-profit companies. Kaizen Institute India (2019) emphasized the systematic process of Kaizen for continuous improvements in the system by eliminating Muda, Mura, and Muri. Farris et al. (2016) laid emphasis on how to achieve specific targets with Kaizen by forming cross-functional teams and departments. Bhuiyan and Baghel (2005) demarcated the word "Kaizen" as an organizational culture, used for eliminating wastes from all areas to sustain and gain profits. They also stated that Kaizen is the only way to create a self-sustained improvement culture in a company. Paul Brunet and New (2003) wrote that Kaizen is a systematic chain of activities with the intended purpose of accomplishing organizational goals. Williams (2001) showed in his research that continuous improvement (Kaizen) methods can reduce the production cost of overall processes in the long run. Fujimoto (1999) clarified that Kaizen is a systematic way to change the organizational behaviour for creating a new culture. Cheser (1998) described Kaizen as trifling but continuous improvements at a workplace to identify and eliminate several wastes and reduce lead-time. Imai (1997) defined Kaizen as a commonsense, low-cost approach to management and enlightened about House of Gemba management that manages overall processes in order to produce cheaper, better, and faster. He emphasized the simplicity of the Kaizen concept that is for everybody irrespective of designation in the company. It should be done every day and everywhere for the effectiveness of Kaizen. Womack and Jones (1996) defined Kaizen as a systematic approach for the identification and elimination of waste.

13.3   KAIZEN ORIGIN AND IMPLEMENTATION

Imai (1986), founder of the term "Kaizen," in his book, Kaizen: The Key to Japan's Competitive Success, defined Kaizen, a Japanese word [改善 (Kaizen)=改 (Kai-Change) + 善(Zen- Better)] as change for better. After the success of the book, many researchers and scholars shifted their attention to "Kaizen." Also, for the first time in year 1993 the *New Shorter Oxford English Dictionary* included the word "Kaizen" and defined it as continuous improvement in working practices, attitudes of people, and efficiency, as a business philosophy. The understanding of Kaizen by different sources given above is based on their learning and experience, but coincidently the gist of all is that Kaizen is "Continuous Improvement" or "Change for Betterment"

Kaizen is a journey, not a destination. The ultimate goal of Kaizen is to identify non-value adding activities that are a waste for an organization and cause low profits or sometimes cause the company sickness. Three basic wastes are: Muda, Muri, and Mura. To improve the performance of the company is not the job of a single person. All employees are equally responsible for improvements, regardless of their designation and roles and responsibilities. The uniqueness of Kaizen is giving equal weight to both Chairman and Doorman in a company. Kaizen is a way of engaging everybody in the process of continuous identification and elimination of Muda, Muri, and Mura, preferably daily.

Kaizen is highly respected in Japan in all areas, and can be seen outside companies, like in schools, railway platforms, airports, markets, and even on roads. Kaizen does process improvement using the lowest investment, use of a new paradigm, observations, speed, and involvement of all stakeholders.

Before implementation of Kaizen, it is necessary to understand its foundation and thumb rules, which are as follows:

- Kaizen always works with the customers' perspective, and creates value for him.
- It is workplace centered and directs users to go to the workplace, see the reality and processes, and make real decisions based on facts and findings.
- It advocates to follow a systematic approach. Do not jump. Be crystal clear at each stage.
- It works on team building, empowering team spirit, and ensuring proper alignment to achieve the goal.
- There is always a scope of improvement as nothing is perfect.
- Set new paradigms to find out the root cause of the problem. The best way is to use WY-WHY analysis.
- Motivate people from each level to participate proactively in problem solving, instead of finding excuses.
- Start with the current situation. Do not wait for perfection.
- Quality production should be the aim; quick action should be taken as mistakes occur.
- Always use mind and time. Try to get the solution with minimum use of money.
- Always use the team approach in problem solving.
- Kaizen doesn't have boundaries.

13.4   KAIZEN TOOLS AND TECHNIQUES

Any single technique or combination of techniques can be used for the successful implementation of Kaizen in an organization. Principally Kaizen is a philosophy adopted for changing the system for betterment or continuous improvement. Tools and techniques of Kaizen are:

1. **5 S:** 5 S is the best tool, propounded by Imai (1986) for ensuring 100 percent participation and engagement of all in an organization. It is well known to change the culture of an organization by changing behaviours, habits, and the work environment to bring overall positive changes. It can be understood in an easy way with the help of the following table and Figure 13.1. (Table 13.1)

A brief explanation of the 5 S is given below:

a. **Seiri/Sorting/Segregation:** The "First S" focuses on the segregation of necessary and unnecessary items at the workplace. Start with a list of all available items at the workplace, and ask the question to yourself whether this item is needed or required. If it is not, attach tags like red tag, yellow tag, purple tag, or green tag (red tag – not needed at all; yellow tag – needed but

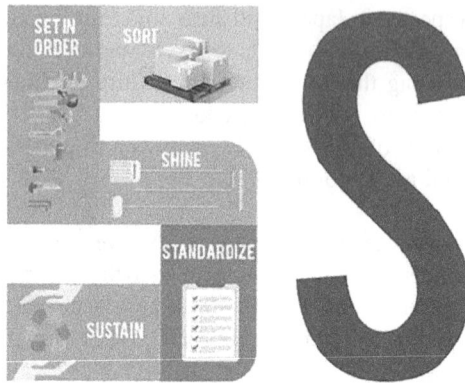

FIGURE 13.1   5 S.

Source: Tyagi K and Tyagi M (2020). A Guidebook on 5 S Through Real Workplace Examples.

TABLE 13.1
5 S Meaning (Japanese to English)

Japanese Word	English Translation
Seiri	Sorting or segregation
Seiton	Systematic arrangement
Seiso	Shine
Seiketsu	Standardize
Shitsuke	Self-discipline, audit

FIGURE 13.2   Sorting (Seiri).

Source: Created by author.

not now; purple tag – needed but not here; green tag – needed) and remove the items from that place. Refer to Figure 13.2; Sorting (Seiri).
b. **Seiton/Systematic arrangement:** The "Second S" advocates the systematic arrangement of all available items. It means that all items must be arranged in a proper way. Use the PEEP policy (a place for everything and

everything in its place). There are some principles for arrangement that help in creating a perfect place to reduce search time and improve retrieval time. This S mainly focuses on what to store, where to store it, and in how much quantity it should be stored.

c. **Seiso/Shine:** The "Third S" emphasizes neatness, cleanliness, and proper arrangement of items and ensures a shiny, clean workplace. In this step the meaning of cleaning is not merely cleaning. The workplace should shine. To keep the workplace/shop floor shined, a detailed cleaning schedule is made, including what to clean, how to clean, when to clean it, and cleaned by whom. The objective of this S is to keep all items in their original operating condition.

d. **Seiketsu/Standardize:** This is the "Fourth S." This S focuses on standardization, and various types of standards are defined, displayed, and taught, on visual boards, with colour codes and fonts. A standard must be created as per the requirements of the workplace, like knowledge, method sheets, visual aids, checklists, standard operating procedures, etc.

e. **Shitsuke/Self-Discipline:** In the "Fifth S," implementation of the above four steps is ensured. The best way to ensure this is an audit. A detailed audit sheet is designed, covering all five steps, and regular audits are conducted. Generally, there are three levels of audit, which are self-audit, audit by departmental head, and audit by third party (external party/consulting company). Per the audit findings, a detailed improvement plan is made and implemented.

5 S Implementation Approach: To ensure successful implementation of 5 S, the following steps must be followed meticulously:

- Appoint one person as the 5 S champion.
- Make a 5 S implementation committee, covering the heads of all functions in the company.
- Take the layout map of the entire company, and divide the company into a sizable area. (These areas will be termed "divisions.") Assign one area to one 5 S steering committee member. These members will be called "divisional champions."
- Identified divisions will be divided further into small areas, which will be called "zones." The divisional champions have to identify one person from each zone and appoint him/her as the zonal champion. Please note all workers working in these zones will be members by default. No exclusion is permitted in 5 S.
- Arrange a training program for all divisional and zonal champions. These champions will further train all members in their zones.

Make the 5 S implementation schedule company-wide. Select one zone to start with to make it a model zone.

- On the kick-off date in a selected zone, meet at the scheduled time (call all divisional and zonal champions), start with Chairman words, conduct the 5 S audit, then start with first "S" activity. Do the first "S" thoroughly, ensure it is done, then subsequently follow all steps as mentioned above.

- Ensure regular audits of 5 S in the model zone (selected initially to start with). The score must be at least more than 80 percent. Also, make sure the internal 5 S-certified auditors are available to conduct audits.
- Design an attractive reward and recognition policy.
- Ensure regular (at least monthly) appreciation of performers. Once the model area is ready, invite all other zone members to learn from this model area. Train them at the workplace again.
- After the above steps, follow the 5 S implementation plan and enjoy success.
- Always be careful about training, leading from the front, involvement of all top leaders during 5 S activities, rewards, and recognition to ensure the success of 5 S.

5 S looks simple but is equally challenging in implementation, so companies should implement it in a very true and honest manner. This one tool only can change the perception. This is very helpful in creating a base for continuous improvement. 5 S must be implemented in a systematic way, many times these five words are also called 5 Steps of 5 S. Some of the **key benefits** of 5 S are:

- Identification of scrap and unused items: Companies many times earn a significant amount of money selling their scrap items. Many times a company finds some hidden valuable items, so it's a cost saving and money maker for company. The money stuck up in scrap is now available to use in some other prioritized item.
- Space freed up: Out of 5 S activities companies, get freed-up space. This is huge profit as they can improve their capacities in the same available space, with no need of additional construction. They can defer the investment on the same.
- Controlled inventory: 5 S set up the principles only to keep standard inventory of items, thus saving on damages, pilferages, and the cost of excess inventory.
- Habit of continuous improvement: The biggest benefit out of 5 S is tasting and developing the good habits that are the base of continuous improvement. After successful 5 S implementation, a high sense of moral engagement and belongingness is achieved in all companies.

2. **PDCA Cycle:** The PDCA Cycle is also one of the core Kaizen techniques, also known as the Deming wheel or Shewhart circle. It was introduced in the 1950s by Dr. William Edwards Deming, inspired by the idea of his mentor, Walter Shewhart. The PDCA cycle shows a clear path to do things in a correct and easy manner. It is shown in Figure 13.3.

Implementation Approach: The PDCA cycle can be implemented with the help of eight steps that must be followed for productivity improvement, quality improvement,

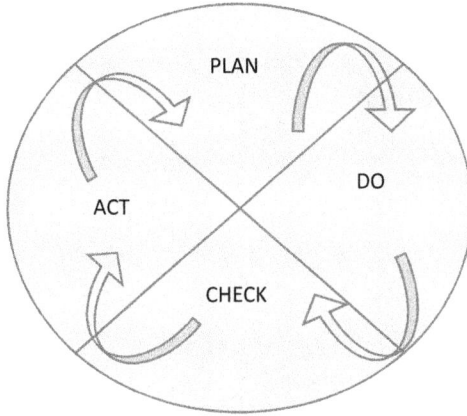

FIGURE 13.3 PDCA Cycle.

Source: Created by author.

TABLE 13.2
PDCA Implementation Steps

Stage	Steps
Plan	1. Identify the problem, preliminary objective, base level, target level, team information, and benefits.
	2. Go to the relevant workplace, collect the data to verify base level, and check the current condition.
	3. Analyze the data. Drill down to the last level. Find out the root cause.
	4. Decide the improvement actions and make an implementation plan.
Do	5. Implement an improvement action plan.
Check	6. Check and verify the results regularly after implementation.
Act	7. If results are in line with the target, prepare new standards, then train and implement users. If not, go back to Step 1.
	8. Make a plan for follow-up and audits and execute accordingly.

downtime reduction, inventory optimization, customer sanctification enhancement, safety improvement, cost optimization, and many more (Table 13.2).

3. **ECRS Technique:** ECRS is one of the effective approaches/techniques of Kaizen for process improvement. It was industrialized for process improvement, and the results were also remarkable in terms of reducing processing time and anticipating efficient working steps that can eliminate unnecessary actions, movement, and waiting time. It is the best technique and must be used when working conditions are related to human works, material cost, and waste material, as well as when working methods are not efficient.

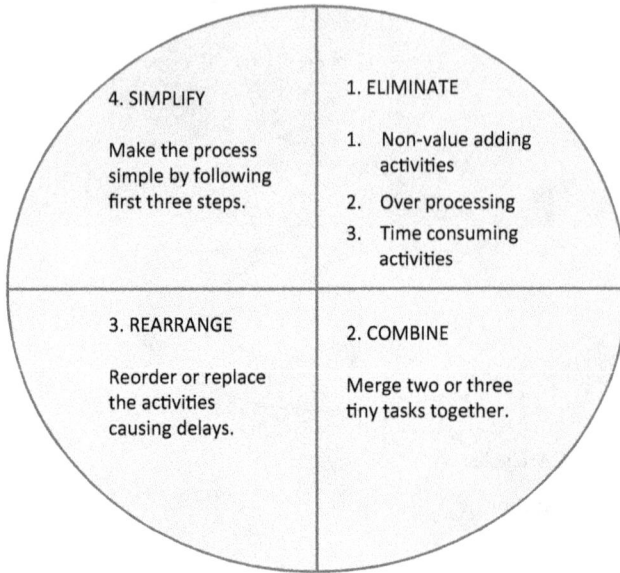

FIGURE 13.4 ECRS.

ECRS represents the four core concepts, shown in Figure 13.4/ECRS stands for the following:

    a. **E: Eliminate** non-value adding activities which are the main cause of waste. These can be seen in manufacturing such as idle time due to uneven allocation of work-on-work stations, needless movement, and unnecessary work steps.
    b. **C: Combine** work steps to eliminate unnecessary work steps that will reduce total processing time.
    c. **R: Rearrange** activities that are the cause of longer production cycle or longer path for material handling process. This step will reduce the distance and number of movements.
    d. **S: Simplify** the methods and introduce new methods that could be done with the help of simple tools, machines or equipment and support operators.

Implementation of ECRS:
- First, draw a process map and write down the steps of the process. These steps should be written on the lower level.
- Write down known information/data about the process steps, such as the following:
  - Time to be taken to complete one activity
  - Show hand offs
  - Elapsed time: Amount of time elapsed between hand over and take over

  4. **SMED:** SMED stands for **S**ingle **M**inute **E**xchange of **D**ies, introduced by *Shigeo Shingo* to reduce the changeover time (change of dies) of pressing

machines. This is a systematic approach, developed to reduce changeover and setup time. The ultimate purpose of this technique was to reduce the changeover time from hours to minutes or less, thus the single minute concept. Although every emplacement cannot be completed in single-digit minutes, ways can be sought to reduce it. The best way to implement SMED is to record the whole activity and setup, study each and every activity, and see the things that can be eliminated.

13.4.1 PROCESS OF IMPLEMENTATION SMED

13.4.1.1 Step 1: Identify the Process

The first step is focusing the area of improvement on reducing changeover time. It is carried out when:

- The changeover is too long and needs to be as short as possible.
- A lot of variances have been noted in the past during changeover times.
- Multiple operations are done frequently.
- Proper training has been given to all employees involved in the changeover process.
- Longer set-up time is a big bottleneck, affecting overall operation adversely, and changes will have an immediate and positive impact.

13.4.1.2 Step 2: Identify Elements

Video recording will help to identify areas of improvement. Each element of the process should be measured by the amount of time taken and the cost. Then, activities done by humans and machines are mapped out.

13.4.1.3 Step 3: Separate External Elements

In the third step, all external mechanisms of the job are parted. On each mechanism, the question considered should be: Can this job be done while the equipment is working or the process is going on? If yes, it can be done before the actual setup or changeover is done.

13.4.1.4 Step 4: Convert External to Internal

In this step, a list is prepared, and those elements that could reduce time and cost significantly in the process are prioritized first. This step requires advanced preparation on many elements. Reconfiguring machinery can also be involved or investment can also be done to arrange extra machinery so that external elements could be handled while the main process is going on.

13.4.1.5 Step 5: Streamline

The final step advocates to make the process simple and easy. That is why in this step, the internal elements are simplified to make the process less time-consuming. Every single element must be studied to find out the weak link taking more time and causing a longer process. This step also suggests to involve new or advanced equipment or a modification in existing equipment to reduce setup time and make the process easier, smoother, and faster. Simultaneously, remove all non-value adding movements and waiting time or idle time.

Value Stream Mapping (VSM): This is a visual display of all critical steps in a specific process that can easily be measured in terms of consuming time and volume at each stage. It is used to show the flow of materials as well as information as they move in a process. Mainly this is used to identify the unnecessary elements in a targeted process.

13.4.1.6 Implementation of VSM

VSM can be implemented easily in any organization. For successful implementation, do the following:

1. The first step is to identify areas for improvement that can be non-value adding activities or more time-taking elements, such as the material handling process, routing, the number or frequency of set-ups, etc.
2. After identifying the activities, collect the necessary information about the targeted elements that can be any one of the above stated elements.
3. After collecting the required information, team members are apportioned specific roles and responsibilities for further study.
4. For proper research and analysis, more and more data and information are collected, and team members add details to value stream maps of the process and conduct time studies of relevant operations.

13.5 RECOMMENDATION

Kaizen is a philosophy that is able to change the culture of any organization and results in a systematic and sincere way. In this chapter, five of the most powerful tools/techniques of Kaizen have been explained along with an implementation approach. By using the same, a company can benefit in the following key areas:

- Optimum utilization of human resources
- Savings of space to produce more with the same facility
- Elimination of non-value adding activities that cause idle time, idle labour, and a longer production cycle
- Brilliant results in all processes
- Enhanced customer satisfaction
- Improved morale of the entire team and a very high sense of belongingness
- A culture of self-sustained improvement

If companies implement the Kaizen philosophy, the above stated techniques will be a boon in the difficult time. Kaizen works on a proactive approach. It will help a culture to work in a proactive manner and avoid a crisis before its occurrence – like when a person is sick, jogging is impossible, but regular exercise can prevent illness.

13.6 CONCLUSION

Kaizen philosophy was adopted and implemented by Toyota first in Japan. Toyota set examples for its competitors and others in the area of utilization of manpower, teamwork, space, and time as well as many more areas that can be causes of high waste or non-value adding activities in an organization. By using simple and very effective Kaizen techniques or methods, with full passion, any company can work like Toyota. The requirement is a high zeal to implement Kaizen. This chapter explains the most powerful Kaizen methods with an implementation approach that can be used with little effort. After COVID-19, things have started to return on the right track again, but if companies want to avoid negative consequences and recover in a swift manner, they should work with the Kaizen methodology. Kaizen will improve the immunity of companies by differentiating costs and waste. Companies can use Kaizen to eliminate all types of waste so that after recovering from a tough time, they will be in a healthy state.

REFERENCES

Bhuiyan, N., & Baghel, A., "An Overview of continuous improvement: From the past to the present", *Management Decision*, Vol. 43, Issue 5, 2005.

Carvalho Cleginaldo Pereira de, & Reis Bruno dos. "The lean manufacturing system applied to an auto parts industry in the heavy vehicles segment", *Global Journal of Engineering and Technology Advances*, Vol. 7, Issue 2, 2021, pp. 037–049. 10.30574/gjeta.2021.7.2.0067

Cheser, R. N., "The Effect of Japanese Kaizen on Employee Motivation in US Manufacturing", *International Journal of Organizational Analysis*, Vol. 6, Issue 3, 1998, pp. 197–212.

Fujimoto, T., *The Evolution of a Manufacturing System at Toyota*. Oxford University Press, NewYork, 1999.

Farris, J. A., Van Aken, E. M., Doolen, T. L., & Worley, J., "Critical Success Factors for Human Resource Outcomes in KAIZEN events: An empirical study, 2016.

Georgise, F. B., & Mindaye, A. T., "Kaizen implementation in industries of Southern Ethiopia: Challenges and feasibility," *Cogent Engineering*, Vol. 7, Issue 1, 2020, pp. 1823157.

Imai, M., *Kaizen: The Key to Japan's Competitive Success*, McGraw-Hill Pub. Co., New York, 1986.

Imai, M., *Gemba Kaizen: A Commonsense, Low-Cost Approach to Management*, McGraw-Hill Pub. Co., New York, 1997.

Iwao, S., "Revisiting the Existing Notion of Continuous Improvement (Kaizen): Literature Review and Field Research of Toyota from a Perspective of Innovation", *Evolutionary and Institutional Economics Review*, Vol. 14, Issue 1, 2017, pp. 1–31.

Kaizen Institute (India) (2019), [online] available from; https://www.kaizen.com/learn-kaizen/learning-materials.html

Paul Brunet, A., & New, S., "Kaizen in Japan: An Empirical Study", *International Journal of Operations & Production Management*, Vol. 23, Issue 12, 2003, pp. 1426–1446.

Resource Outcomes in Kaizen Events: An Empirical Study", *International Journal Production Economics*, Vol. 117, Issue 1, 2009.

Sehleanu, M., & Flore, E. S. (2019). Continuous Improvement Through Kaizen Management System: A Case Study. In Proceedings of the International Management Conference, Faculty of Management, Academy of Economic Studies, Bucharest, Romania (Vol. 13, No. 1, pp. 25–36).

Tyagi, M., Singh, M. N., Gautam, R., & Sharma, M. R., "Adopting the Philosophy of Kaizen "To Cut Waste Not Cost" in Difficult Economic Scenario-Covid-19," *Solid State Technology*, Vol. 63, Issue 5, 2020, pp. 1277–1289.

Tyagi, K., & Tyagi, M. (2020). A Guidebook on 5S Through Real Workplace Examples.

Womack, J. P., & Jones, D. T., *Lean Thinking: Banish Waste and Create Wealth in Your Corporation*, Simon & Schuster, New York, 1996.

Williams, M., "Maximum Cost Reduction Minimum Effort," *Manufacturing Engineer*, Vol. 80, Issue 4, 2001, pp. 179–182.

14 Potential Electricity Production of Roof-Mounted Solar PV Systems in a Row House Area in Sweden

Alireza Bahrami, Arman Ameen,
Mattias Halvarsson, and Mikael Aho
Department of Building Engineering, Energy Systems and
Sustainability Science, Faculty of Engineering and
Sustainable Development, University of Gävle,
Gävle, Sweden

14.1 INTRODUCTION

Globally, there is a steadily increasing demand for energy, which is mainly produced with fossil fuels. The earth's atmosphere is negatively affected when large amounts of greenhouse gases (GHGs) and other pollutants are released with burning the fossil fuels. The availability of these non-renewable fuels is finite, and therefore the eyes are increasingly directed towards renewable energy sources such as solar energy [1].

Sweden's energy policy goal is to have 100% renewable electricity production by the year 2040, and to achieve that, all available alternative energy sources need to be developed and implemented in the Swedish electricity mix. Continued increased population contributes to increasing electricity demand, which is one of the major challenges that, together with the prevailing climate changes, must be dealt with. The share of solar energy in the Swedish electricity mix is currently very small, but in recent years it has become an increasingly economically profitable source of renewable energy, partly because technological development has contributed to cheaper and more efficient solar cells, and partly because of financial subsidies from the state to install the solar cells. Currently, solar PV systems account for only 0.2–0.4% of Swedish electricity production, but the government aims to increase the production to 5–10% by 2040 [2,3]. Continuous development is underway to enable cost-effective systems that both meet the energy demand and deal with challenges such as the system's influencing parameters, which often look different for each new installation site [4].

DOI: 10.1201/9781003365525-14

The technique of placing solar cell modules on roofs or alternatively along vertical structures on buildings and converting the sun's radiation into electricity has been used since the late 1990s. Basically, the technology works in such a way that free electrons are produced with the help of semiconductors by capturing the photons of light and generating varying effects depending on the design of panels in terms of manufacturing material and size [4]. Based on the choice of raw material and manufacturing method, panels can either be semi-transparent or non-transparent and intended to cover building's roofs, walls or integrated into windows. However, for a successful design of a solar cell system in an urban environment, several key factors need to be met, partly an estimation of potential unshaded and free roof surface for installation, and partly a measurement of incident radiation. Defaix et al. [5] studied PV system installation on building structures in EU countries. The results indicated that the potential exists to produce 840 TWh of solar electricity for the EU, which would correspond to more than 22% of the predicted annual electricity demand in the EU in 2030.

Thirty-four countries in the EU were evaluated by Raghoebarsing et al. [6], and it was shown that they produced a total of 120 TWh of PV electricity in 2018. Germany produced the highest amount of 45 TWh and had the highest share (8.4%) of PV in its electricity production. It was also concluded that several countries had significant unused PV potentials, such as Israel, Turkey, Portugal, Spain, and Cyprus. The possibility of deploying PV systems in Seoul buildings was evaluated by Byrne et al. [7]. It was found that nearly 30% of the city's total annual electricity demand could be provided by roof PV panels.

Smaller systems with smaller producers historically have several advantages regarding both the production and sale of electricity to the grid [3]. For larger installations on buildings concerning multi-apartment buildings, there is great potential; however, it cannot yet be fully exploited. An obstacle to such installation is that the solar cell systems must be connected to the electricity grid via the property's electrical system with fuses smaller than 100 A, which means that many larger conceivable installations fall away.

Furthermore, there are limitations regarding how and by whom the self-produced electricity is used [3]. For example, an organization or a tenant-owner's association does not receive any tax exemptions if the produced electricity is spread over several houses, even if all buildings are owned by the organization itself and the electricity never leaves the buildings. In other words, the limit for how the produced electricity can be used is rather determined by the layout of the electricity metres than the ownership of the solar cell plant. In the study conducted by Warneryd and Karltorp [3] on three tenant owner's associations where they tried to crystallize a picture of how their expectations of solar cell systems were met. It turned out that the financial expectations were not met. Instead, the projects brought with them synergy effects that grew increasingly strong during the work and proved to be advantages for the solar systems. Primarily, one of the effects was that the owners adopted a more sustainable and environmentally conscious attitude, which contributed to the choice of installing solar panels.

All the world's cities currently only take up about 2% of the earth's surface, but in 2007, an estimated half of the world's population lived in cities and accounted for about 3/4 of the use of the earth's resources. Forecasts illustrate that by the year 2035 about 61% of all people will live in cities [8]. Cities in many cases possess characteristics that are decisive for the potential of installing solar cell systems, i.e., many buildings have the potential of mounting panels on an often-unused roof surface, which is also often relatively shadow-free. This means that there is potential for the utilization of solar cell technology whose properties are interesting in increasingly energy-intensive cities and city centres [4]. In densely built-up areas, there are large areas of roof surface with potential for installing PV systems. Such surfaces have been examined in previous studies to assess their conditions for mounting PV systems. A study in Toronto, Canada, found that with a roof PV system, Ryerson University was able to meet 19% of its electricity demand during peak consumption hours [1].

Residential buildings have great potential for installing PV systems and 2/3 of the installed systems in Sweden today are mounted on building shells [3]. Minuto et al. [9] carried out a study on a residential area in Italy where they looked more closely at how well a roof solar cell system would work. They stated that there is potential, even if in this case it could only cover about half of the electricity that the property required, partly because of limited roof space and partly because panels were not installed in the north direction.

Yang et al. [2] estimated that the potential roof area for solar cells in the city of Västerås, Sweden, is roughly the same size as in the city of Gävle, Sweden, about 5.74 km^2. The total capacity of these PV systems then, based on three different scenarios, was 727, 848, and 956 MW_p, with respective annual electricity productions of 626, 720, and 801 GWh. As PV systems are installed by private individuals and small companies, this also can contribute to a decentralization of electricity production [3].

The solar cell market took off in Sweden in 2005 when a targeted financial effort in the form of a capital investment system was introduced. Through the design of the investment system with a lower proportion of upfront costs during the design phase, the profitability of systems in the housing sector could be increased and has thus enabled the sharp increase in the number of systems in operation for houses, and it is one of the reasons why the solar cell systems in Sweden are predominantly decentralized in nature [10].

The economic aspect of mounting solar cells was reported by Haegermark et al. [11] in a study done in the city of Gothenburg, Sweden. They reflected the importance of financial support for solar cell systems that exist in Sweden, which can generally be divided into two parts; partly for smaller systems of private customers in which there is a tax exemption for the sale of excess electricity to the grid, and partly for slightly larger systems of companies, which often receive subsidies during design. Notably, only 10% of the examined objects were considered economically justifiable without subsidies or tax exemptions. A varying proportion of the systems were then considered to be able to be implemented with good finances thanks to subsidies or tax exemptions depending on their size. Nine out of ten systems would economically be feasible if both financial reliefs were applied to the systems. It should be noted that even if the economic picture were more complex than that, the importance of financial support to keep solar cell production attractive for residential areas would still be highlighted.

In recent years, as an increasing number of systems has been installed, and with a technology that has now begun to establish itself, a reduction in costs for both purchase and installation can be seen, resulting in the subsidy system that has been in place since 2005 and was abolished in 2020 [10]. Since the old subsidy system, in the form of capital support that existed for the last ten years, was abolished, no new valid investment subsidies have been added. However, what a private individual can do is use tax deductions when installing the systems. Tenant-owner's associations are entitled to a VAT deduction for the installation of solar cell systems corresponding to the excess electricity that is intended to be sold online when the system is in operation. For financing solar cell systems, there are also special loans dedicated to the investment of solar energy for private individuals. The larger banks are those that offer so-called solar cell loans of up to 350,000 SEK with a repayment period of up to 12 years and at an interest rate of up to 2.9%. Every year, the International Energy Agency (IEA) compiles a comprehensive analysis in report form regarding the solar cell industry [8].

Until July 2021, self-produced electricity was tax-exempt, provided that it was used in the building where it was produced and the installed solar power did not exceed 225 kW. This limit has now been raised to 500 kW, but the Solar Commission is pushing for that limit to be removed completely, which has already been done in many other EU countries. If this is done, it is believed that the potential opens for solar cell installation on about 8000 buildings, which could produce about 2 TWh electricity. The Solar Commission's goal for 2030 is 30 TWh electricity to be produced in Sweden, of which 20 TWh on roofs and 10 TWh in solar parks. Moreover, it is said that Sweden should set long-term goals and adopt a similar strategy for the solar park as for wind power. Total installed solar cells in Sweden in 2020 were just over 1000 MW, a market that has grown by 50–80% every year since 2014 and can be divided into various sectors such as centralized, industrial, commercial, and residential systems [8].

As limitations already exist with Sweden's seasons and varying hours of sunshine throughout the year, it is of great importance to have correctly mounted panels to achieve good function. Correct installation of the panels in terms of direction and above all slope, together with consideration of the surrounding shading, are decisive factors for the success of an efficient system. Many studies have been done on planning the establishment of solar cell systems with the consideration of the placement of the panels to achieve optimal direction and slope, and what optimal slope and direction really mean. The system is complex and contains several parts that need to be managed in the right way for the system to deliver the expected electricity output and have economic acceptance [12]. Theoretically, optimal angles can only be used as a starting point to then identify the practically optimal angle for the installation location. Working only with the theoretically optimal angles and slopes risks contributing to unprofitable solar cell system designs. As Yadav and Chandel [12] highlighted, this is due to, among other issues, the geographical location where, for example, the Nordic countries have different conditions with a higher degree of hours when the sky is covered by clouds and the angle of incidence of solar radiation. However, several research works agree that even if generalizing optimal guidelines are followed, the result is that the greatest importance in each new location must be placed on individual conditions such as the form of the roof, geographical location, and the location's shading and surroundings [4,12,13].

Solar panels are very sensitive to shading in several respects, which must be taken into account in the calculations. Shading can occur from chimneys, ventilation shafts, and the like, as well as from surrounding buildings and forests. Even partial shading has a large impact on the efficiency of the PV system and is a significant loss factor to consider [14]. Partial shading also risks shortening the working life of the cells, as this type of shading is harmful to them, unlike total shading, which only affects annual electricity production [15]. The cells in most solar modules are generally connected in a series, but by dividing the cells into smaller groups, the losses in partial shading can be reduced [14]. In an evaluation of shading of different types of solar cell modules, it was found that, with the correct distribution of cells, losses can be reduced by 30–40% [15]. Considering this point, simulations with different solar cell types should be done to find the type that fits best in the current research.

In addition, there is also a risk of so-called self-shading if the solar cells are not mounted correctly. To get the maximum effect, especially on flat roofs, the panels must be angled, but they risk shading each other when the sun is low. The problem can be avoided by determining a longer row distance between the panels to achieve an optimal distance without self-shading, especially during the winter half of the year. However, a larger row spacing means that fewer panels can be mounted, which reduces the installed solar power, but in return, the produced electricity can be increased per surface area. The row spacing (PF) is calculated using the following equations:

$$PF = \frac{1}{d} = \left(cos\beta + \frac{sin\beta}{tan\alpha} \, cos \, \gamma \right)^{-1} \tag{14.1}$$

$$cos\alpha = sin\delta sin\Phi + cos\delta cos\Phi cos\omega \tag{14.2}$$

where β is the slope of the panel, α is the sun's altitude defined by equation (14.2), γ is the sun's azimuth, δ is the soil's slope, and ω is the hour angle. Additionally, a balance between optimal slope and optimal row spacing gives the highest electricity production per m^2. Thus, scenario one with an optimal slope on the panels gives the highest kWh/kW$_p$, while scenario two, with a trade-off between slope and row spacing, gives a higher proportion of kWh/year. In the third scenario, the panels are mounted in the east/west direction to utilize a greater proportion of solar radiation during the early and late hours of the day. A higher installed solar power is required for this scenario but gives the highest proportion of kWh/year [2].

14.2 MODELLING

The computer software IDA ICE 5.0 beta was used in this study. This software is a building performance simulation software developed by EQUA that supports multiple or single zones to simulate thermal comfort and moist transfer, and also to evaluate energy loads and usage as well as utilization of different energy sources for building integrations such as solar power, wind power, or different types of heat pumps. The simulation can be done on room level or for an entire building, which was conducted in this study. The software has been validated according to CEN standards EN 15255–2007, 15265–2007, and 13791, as well as ASHRAE Standard 140–2004. The software has also been validated in the present study [16].

IDA ICE simulation software is widely used in Nordic countries and in Europe for simulating energy performance estimations [17,18], and for comparison of energy performance in different climate regions [19].

In this research, measurements of the height and length of the houses, as well as the distance to the surrounding forest, were carried out on site with a laser metre (Hilti PD 30). The slope of the roof was calculated as 14.5° by measuring the width of the buildings as well as the highest and lowest points of the roofs. All the roofs in the area have the same slope. Drawings of the area were assigned by the client. Climate data that exist in the software were utilized. As the climate data for Gävle were missing in the database, data for Söderhamn that is only six miles away from Gävle were used. A map image from Google Maps was then processed in an image editing program to remove garage lines, garbage rooms, and other buildings whose roofs in the study were not relevant for the installation of the solar cells. The map image was later employed as a basis for the layout of the buildings and the surrounding forest, which is illustrated in Figure 14.1.

In the next step, the original image from Google Maps was imported as a bitmap in "floorplan" in the software (Figure 14.2), and the edited area map from Google Maps (Figure 14.1) was then used as a basis for creating the models. Figure 14.2 was utilized as a basis for how the buildings would relate to geographical direction and to each other. Different zones defined the area of each building where the buildings were modelled using metre scale coordinates.

Figure 14.2 Original image from Google Maps as a bitmap in "floorplan" view.

In relation to the deployment of the buildings, some of them were adjusted geometrically to simplify the work to some extent. The adjustments were, for example, if two houses were close to each other but slightly offset, these buildings were merged into a single section to be able to put a solar cell system over the two roofs. This was a simplification prior to the simulation, which in a small test showed small differences in the results and is discussed in Section 14.3 of the current chapter. The simplification was not expected to affect the results significantly as the direction of the solar cell modules only changed marginally, and the size and shadow image of the buildings did not change at all. The houses were thereafter divided into two categories, south and east/west directions, which were colour-marked. The roofs in the north direction were not dealt with in the simulations as they were not expected to be able to produce enough electricity to be profitable to mount solar cells on the roofs. The area consisted of 90 terraced houses located in different directions. The houses were distributed in a simplified model with 49 blocks in the east/west direction and 41 blocks in the south direction. Models were finally created in the form of 8 blocks in the east/west direction and 7 blocks in the south direction, as demonstrated in Figure 14.3.

FIGURE 14.1 Map of examined area.

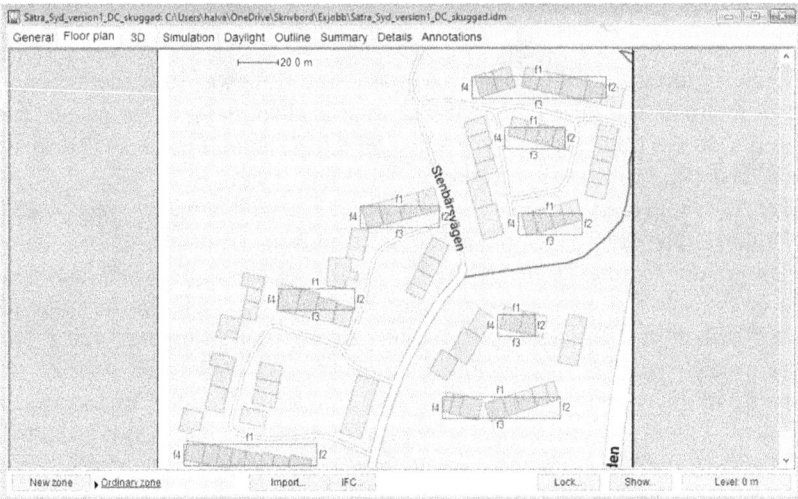

FIGURE 14.2 Building layout used in IDA ICE software.

FIGURE 14.3 Simplified and colour-marked model of area (arrows illustrate distance to forest).

The next step was of great importance as the ceiling was to be defined. In the area, all the roofs had the same slope and were pitched roofs. In the pitched roofs, both sides of the ridge were of the same area and angle. Figure 14.4 displays a typical modelled roof in the software.

Then, the setup of the solar cell system was done, and PV panels were placed on the buildings. The solar panel model was selected from the database. The deployed panels were divided into strings where the number of panels per string was chosen based on the selected inverter's electricity. Inverters were selected from the database for each string. The defined PV strings are depicted in Figure 14.5.

The panels were placed partly directly on the roofs, which had a slope of 14.5°, and partly in the optimal slope of 45°. According to the Swedish Energy Agency, placement of the solar panels directly on the roof is recommended as it usually does not require a building permit and it results in a simpler system design, which in turn results in lower installation costs. In addition, it is stated that even if the roof slope deviates from the optimal slope of 30–50°, it has little impact on the annual

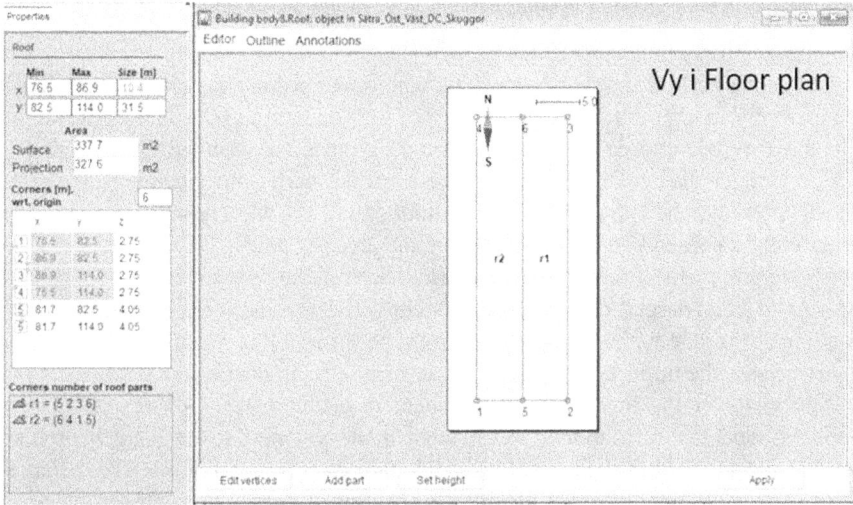

FIGURE 14.4   Modelling of roof in software.

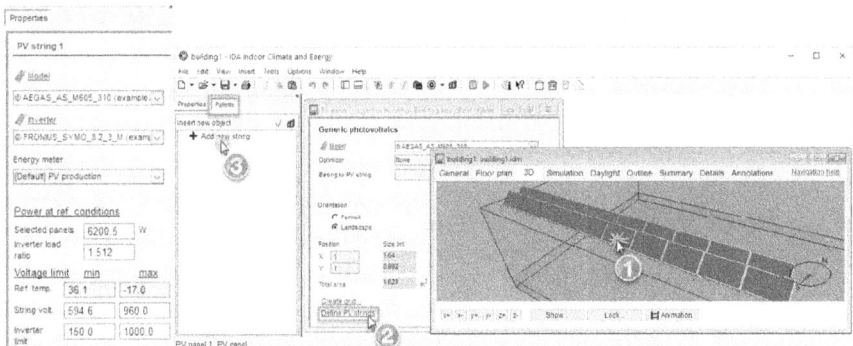

FIGURE 14.5   Defining PV strings in software.

production, where a deviation of 10° from the optimal slope leads to a reduction of the annual production of only 1–2%.

Only solar cells of the monocrystalline type were used in the simulations as this is the currently market-dominant variant. From the database in IDA ICE, AEGAS AS M605 310 W was chosen as solar cell, Fronius Symo 8.2 M as inverter, and Tigo TS4 L as optimizer; however, it could also be possible to add other components to the database by oneself. The panel with an output of 310 W_p was chosen arbitrarily, as it is currently a normal panel for both output and area per panel. The choice of inverter was governed by the size of the system, as a relatively large inverter was preferred to be able to build larger strings in the systems. The models were simulated in two configurations, one with added shading from the surrounding forest and one without, to investigate the impact of shading on the electricity production.

Site visits of the area were done where measurements of distances between buildings and the surrounding forest were made in order to be able to produce average distances from the forest to buildings for each row of houses. The distances were used as a reference for where the shadow screens, representing the trees, should be placed. The shading from the surrounding forest was created in the software. As all the forest around the area was old, an estimation of the average height of 20 m and width of 0.4 m was made. The exact values were not possible to create due to the varying size of trees, and the fact that foliage, branches, etc., could not be included in the model. In this study, only the electricity production was of interest; therefore, all the properties and loads of the buildings such as lighting, personal loads, heating, and ventilation were set to zero or completely removed, so that unnecessary data were not included in the simulation and in the results.

After completing the simulation, the results were obtained as total produced kWh and kWh per month. The total installed solar power was calculated by hand. The values such as kWh/kW_p and kWh/m^2 could be calculated for the simulated system. When all the models were simulated, analyses and comparisons were performed between the shaded and unshaded systems to find out how they were affected with the different properties.

As data regarding the energy use of houses were not available, general information on the energy use in Sweden was applied as a basis for calculations. The calculations were based on the general data compiled by statistics from [20]. The electricity consumption in a normal-sized villa in Sweden is estimated to be 120 kWh/m^2 per year if it is heated with direct electricity and 40 kWh/m^2 if it is heated with, for example, district heating [20]. The residence in question uses an exhaust air heat pump for heating, which gives an energy consumption of 64 kWh/m^2, according to the manufacturing data sheet.

14.3 RESULTS AND DISCUSSION

Simulations with a smaller module consisting of ten panels with a total installed solar power of 3.1 kW_p were initially conducted to evaluate possible differences between results from the simplified model and from models with modules oriented in the direction according to the drawings. The results revealed differences of less than 1% when simulating the modules directed to the south compared with those

TABLE 14.1
Simulations with Test Modules Directed in Different Directions

Rotation	0°	+10°	−10°	+15°	−15°	Unit
South 0°	2770	2765	2762	2760	2755	kWh/year
Difference	–	0.18	0.29	0.36	0.54	%
East 90°	2386	2319	2453	2286	2485	kWh/year
Difference	–	2.81	2.73	4.19	3.98	%
West 270°	2406	2338	2473	2305	2504	kWh/year
Difference	–	2.83	2.71	4.20	3.91	%

rotated +/- 10–15°, and differences of around 2.8% with the modules placed in the east/west direction (Table 14.1). Of the buildings that are in the east/west direction, there is roughly the same proportion twisted +10–15° as −10–15°, which means that the deviations basically cancelled each other out. As far as differences in these comparisons in IDA ICE were concerned, they were considered negligible, where the simulations were done with the panels oriented in the south and east/west directions.

Validation of the results from IDA ICE was performed in PVgis and WINsun, where two scenarios were simulated in all three software programs with the following setup. All three systems had the installed solar power of 3.1 kW$_p$, and the solar cells were directed towards the south. In IDA ICE, built-in climate data files, the defined solar panel (AEGAS AS M605 310 W), the defined number of panels, and the panel area were used. In PVgis, from its own climate file database PVGIS-SARAH2, a general crystalline silicon solar panel was utilized, and a defined system loss of 14% was preselected. In Winsun, the panels were defined with efficiency, installed solar power (kW$_p$), and solar cell area, which was calculated based on the selected panel in IDA ICE. The solar cells were placed partly in the roof slope of 14.5° and partly in 45°. The reason why 45° was chosen was because it was the angle that the Swedish Energy Agency uses in its comparison between different solar cell models. When comparing the results obtained from the three software, the results from PVgis and WINsun were reproduced with the difference of between 0.2% and 10.8% compared with those from IDA ICE. These comparisons are indicated in Table 14.2, which demonstrates that the results from IDA-ICE are within the acceptable range.

As an important part of the simulations, comparisons were made with the roof's slope of 14.5° against 45° in Table 14.2 to investigate whether it would be relevant to angle the panels up, to further utilize the solar radiation. Based on the obtained results, the slope of 45° would produce about 2.5% more energy than installing the system in the slope of the roof, which means that it is not considered profitable to install the panels in a slope other than that of the roof. In accordance with the Swedish Energy Agency, it is customary to mount solar panels directly on the roof regardless of the slope as it was mentioned in Section 14.2 of the current chapter, in this way, it does not require a building permit, which means having lower installation costs.

TABLE 14.2

Comparisons of Results from Different Computer Software

Software	Slope	Direction	kWh/year	Difference with IDA ICE	
				(kWh)	%
IDA ICE 5.0	14.5°	0°	2770	–	–
PVgis	14.5°	0°	2540	230	8.3
WINsun	14.5°	0°	2470	300	10.8
IDA ICE 5.0	45°	0°	2840	–	–
PVgis	45°	0°	2834	6	0.2
WINsun	45°	0°	2608	232	8.1

The entire residential area was modelled in IDA ICE, where solar cells were placed on all the roofs, except for those facing north as they were not expected to produce enough electricity to be profitable. All the models used solar cells with the electricity output of 310 W_p, optimizers (Tigo TS4 L (DC)), and inverters (Fronius Symo 8.2 3 M) with a nominal electricity output of 8.2 kW. To facilitate the simulations, the area was divided into two models, one with the roofs in the south direction and one with the roofs that have the east/west direction. Two versions were created for each model, one with the surrounding forest deployed and one without, to assess the impact of shading that the forest can have. Table 14.3 lists the results. It is worth mentioning that the installed W_p/m^2 was 190.5. As expected, lower electricity was produced by the shaded models. The values of kWh/year were obtained from the simulation results, while the other values in the table were calculated by hand.

Since several models were built to enable comparisons of how shading from surrounding vegetation affects the potential of the system, it was possible to calculate the difference. Considering the system as a whole, 4.8% more energy was produced in the completely unshaded models compared with those with the surrounding vegetation. Based on the results, the shading did not significantly affect the total electricity production.

TABLE 14.3

Results from Simulations with Solar Cells Placed on the Roof with the Slope of 14.5°

Conditions	kWh/year	kWh/kW$_p$	kWh/m^2	Area (m^2)
South without shadow	214356	1011	192.6	1113
South with shadow	203789	961	183.1	1113
East/West without shadow	367588	869	165.7	2219
East/West with shadow	350164	828	157.8	2219
Total without shadow	581944	917	174.7	3332
Total with shadow	553953	873	166.3	3332

TABLE 14.4

Produced Electricity in Summer and Winter Halves of the Year

Conditions	October-November	December-January	February-March	April-June	June-July	August-September	Unit
South	13397	4605	20798	59516	62906	42567	kWh
East/West	16868	2081	31841	106475	118529	74733	kWh
Total	30265	6686	52639	165991	181435	117300	kWh
	Winter half of year			Summer half of year			
Total	89590			464726			kWh

The total residential area was 8131 m^2, which gave an energy requirement of about 522 MWh/year. Thus, the energy requirement was approximately as large as the total electricity produced 554 MWh/year. However, as it is well known, the need is greatest in the winter half of the year, while the production is greatest in the summer half of the year. Compilation of the monthly electricity production reveals the potential electricity production of about 90 MWh from October to March and about 465 MWh from April to September. Table 14.4 summarizes the produced electricity in the summer and winter halves of the year. As the results show, the electricity production is at its highest level during the summer half of the year and at its lowest level during the winter half of the year. The buildings' energy needs have the opposite nature. Even if the production over the year matches the current need, the skewed distribution of the electricity production means that not all the electricity can be used during the summer half of the year and that a large deficit of the electricity production occurs during the winter half of the year. In order to be able to use all the electricity produced during the summer half of the year, some form of storage needs to be introduced.

As the system looks now, without the possibility of storage, there is, however, the possibility that a relatively large proportion of surplus electricity could be sold to the electricity grid during the summer half of the year, which can be bought back during the winter half of the year, when the energy demand is greater. Currently, this option exists for private individuals, associations, and smaller companies, and then a tax reduction of 60 öre/kWh applies to the exported electricity, provided that the export does not exceed the consumed share of electricity; however, the consumption may amount to a maximum of 30,000 kWh and that the fuse size does not exceed 100 A. Consequently, a balance should be considered for the size of the system to achieve a balance during the summer half of the year, when the system is designed according to the area's energy needs during the summer half year, or to maximize the system's electricity production and export surplus electricity to the grid.

In order to get an estimation of costs and taxes, statistics from the IEA's report were taken. Since 01/01/2021, new tax rules have been applied to the consumption of self-produced electricity, where the ceiling for the installed solar power has been raised to 500 kW$_p$, which means that the tax is 0.005 SEK/kWh. The installed solar power in the model of this study was 634 kW$_p$, which exceeded the electricity limit,

TABLE 14.5
Number of Panels and Installed Solar Power

Model	Number of Panels	Installed Solar Power (kW$_p$)
South	684	212.04
East/West	1370	424.70
Total	2054	636.74

TABLE 14.6
Estimated Cost of System

Cost of System (SEK)	SEK/kW$_p$	SEK/m^2	SEK/kWh
6749444	10600	2025	11.71

but the system could be divided into several smaller systems of maximum 255 kW$_p$ each [8].

To exemplify the cost of the system, the full-scale simulated system was divided into several subsystems of maximum 255 kW$_p$, which gave a total of installed solar power over 500 kW$_p$. The results are reported in Table 14.5. Then, an estimation of the system cost was made according to Table 14.6, in which the cost of 10.6 SEK/W$_p$ was used which resulted in the system cost of 6,749,444 SEK. The cost per installed square metre of solar panels was 2025 SEK, and the system cost based on the electricity produced over a year was 11.71 SEK/kWh.

The payback period for the investment in solar cells in Sweden is on average 10–15 years, and if the payback period in this case were set to 10 years, it would give the system cost of 1.17 SEK/kWh.

14.4   CONCLUSIONS

In the Stebnär tenant-owner's association, each residence has its own electricity metre and bill, which leads to certain tax technical problems if solar cell systems are installed. As the regulation looks in Sweden today, in that case, the self-produced solar energy can only be used in common areas such as meeting rooms, lighting, laundry rooms, garbage rooms, etc. If the self-produced electricity, with the current system, is to be used by the tenant, the property owner is counted as a producer and actor in the electricity market, which means that the property owner is subject to the same laws and tax regulations as electricity suppliers, which makes the situation complex. To get around this, the association needs to adopt the model with one electricity metre per row house length or area and a joint contract with the network supplier, which means that the electricity must be included in the rent. Alternatively, it is up to the property owner whether the tenant must pay a fixed price regardless of the electricity consumption or the rent must instead be adjusted monthly according to how much electricity they use. In such a case, the tenant can use the self-produced electricity from the solar cell system without the property owner being counted as an electricity producer. Thus, even if there is potential of producing own's electricity, this is something that needs to be managed to achieve greater profitability with the system. The latest version of IDA ICE, IDA ICE 5.0 beta, which was used in this study with the new functions for solar panels, has proven to be a useful software for modelling and simulating solar cell systems, and has good potential for comprehensive simulations where the thermal properties of buildings are included in the simulations for the building's total energy demand and own's energy production. Results have shown that there is potential to produce approximately 554 MWh/year, of which 89.6 MWh/year is in the winter half of the year and 464.7 MWh/year is in the summer half of the year, with the solar cell modules chosen in these models. Small differences in the electricity production could be demonstrated when comparing the slope of the panels; thus, it was suggested to place them directly on the roofs. The shading also did not significantly affect the system as a whole. The approximate system cost was calculated at about 6.7 million SEK, and with a payback period of 10 years, the cost of the electricity produced would be 1.71 SEK/kWh.

This study demonstrated that it is possible to utilize unused roofs part of buildings in order to transit to a more sustainable energy production and energy usage. The transition from fossil fuel-based energy sources to clean energy sources such as solar power is crucial to reduce global warming and energy dependency on fossil fuel for electricity production.

REFERENCES

1. A. Chow, S. Li, A.S. Fung, Modeling urban solar energy with high spatiotemporal resolution: A case study in Toronto, Canada, *International Journal of Green Energy*. 13 (2016). 10.1080/15435075.2016.1170686.
2. Y. Yang, P.E. Campana, B. Stridh, J. Yan, Potential analysis of roof-mounted solar photovoltaics in Sweden, *Applied Energy*. 279 (2020). 10.1016/j.apenergy.2020.115786.
3. M. Warneryd, K. Karltorp, The role of values for niche expansion: The case of solar photovoltaics on large buildings in Sweden, *Energy, Sustainability and Society*. 10 (2020). 10.1186/s13705-020-0239-7.
4. A.A.A. Gassar, S.H. Cha, Review of geographic information systems-based rooftop solar photovoltaic potential estimation approaches at urban scales, *Applied Energy*. 291 (2021). 10.1016/j.apenergy.2021.116817.
5. P.R. Defaix, W.G.J.H.M. van Sark, E. Worrell, E. de Visser, Technical potential for photovoltaics on buildings in the EU-27, *Solar Energy*. 86 (2012). 10.1016/j.solener. 2012.06.007.
6. A. Raghoebarsing, I. Farkas, D. Atsu, S. Boddaert, D. Moser, E. Shirazi, A. Reinders, The status of implementation of photovoltaic systems and its influencing factors in European countries, *Progress in Photovoltaics: Research and Applications*. (2022).
7. J. Byrne, J. Taminiau, L. Kurdgelashvili, K.N. Kim, A review of the solar city concept and methods to assess rooftop solar electric potential, with an illustrative application to the city of Seoul, *Renewable and Sustainable Energy Reviews*. 41 (2015). 10.1016/j.rser.2014.08.023.
8. J. Lindahl, A.O. Westerberg, K. Vanky, *National survey report of PV power applications in Sweden 2020*, Swedish Energy Agency and International Energy Agency: Sweden. (2020) 1–86.

9. F.D. Minuto, P. Lazzeroni, R. Borchiellini, S. Olivero, L. Bottaccioli, A. Lanzini, Modeling technology retrofit scenarios for the conversion of condominium into an energy community: An Italian case study, *Journal of Cleaner Production*. 282 (2021). 10.1016/j.jclepro.2020.124536.

10. J. Lindahl, D. Lingfors, Å. Elmqvist, I. Mignon, Economic analysis of the early market of centralized photovoltaic parks in Sweden, *Renew Energy*. 185 (2022). 10.1016/j.renene.2021.12.081.

11. M. Haegermark, P. Kovacs, J.O. Dalenbäck, Economic feasibility of solar photovoltaic rooftop systems in a complex setting: A Swedish case study, *Energy*. 127 (2017). 10.1016/j.energy.2016.12.121.

12. A.K. Yadav, S.S. Chandel, Tilt angle optimization to maximize incident solar radiation: A review, *Renewable and Sustainable Energy Reviews*. 23 (2013). 10.1016/j.rser.2013.02.027.

13. M. Oh, J.Y. Kim, B. Kim, C.Y. Yun, C.K. Kim, Y.H. Kang, H.G. Kim, Tolerance angle concept and formula for practical optimal orientation of photovoltaic panels, *Renew Energy*. 167 (2021). 10.1016/j.renene.2020.11.096.

14. E.D. Chepp, F.P. Gasparin, A. Krenzinger, Accuracy investigation in the modeling of partially shaded photovoltaic systems, *Solar Energy*. 223 (2021). 10.1016/j.solener.2021.05.061.

15. K. Brecl, M. Bokalič, M. Topič, Annual energy losses due to partial shading in PV modules with cut wafer-based Si solar cells, *Renew Energy*. 168 (2021). 10.1016/j.renene.2020.12.059.

16. M. Woloszyn, C. Rode, Tools for performance simulation of heat, air and moisture conditions of whole buildings, in: *Building Simulation*, 2008: pp. 5–24.

17. A. Kabanshi, A. Ameen, A. Hayati, B. Yang, Cooling energy simulation and analysis of an intermittent ventilation strategy under different climates, *Energy*. 156 (2018).

18. S.A. Moghaddam, M. Mattsson, A. Ameen, J. Akander, M. Gameiro Da Silva, N. Simões, Low-emissivity window films as an energy retrofit option for a historical stone building in cold climate, *Energies (Basel)*. 14 (2021) 7584.

19. A. Ameen, M. Cehlin, Reducing energy usage in multi-family housing, in: IOP Conf Ser Earth Environ Sci, 2019.

20. Bergmark W., Normal energy consumption for a villa? *HemSol AB*. (2022). https://hemsol.se/solceller/elforbrukning-villa/ (accessed August 30, 2022).

15 The Double-Edged Sword of AI and Industrial Revolution 4.0

Will They Widen or Bridge the Global Economic Divide?

Ashutosh Sharma and Reeta Rautela
Assistant Professor, School of Liberal Arts, Uttaranchal
University, Dehradun, India

Shravan Kumar
Professor & Dean School of Liberal Arts, Uttaranchal
University, Dehradun, India

Lakshya Thakur
BA First year School of Liberal Arts, Uttaranchal University,
Dehradun, India

15.1 INTRODUCTION

The Fourth Industrial Revolution (also known as Industry 4.0)[1] refers to the current era of technological advancement, characterized by the integration of digital, physical, and biological systems. Industry 4.0 builds on the technologies of the previous three industrial revolutions (mechanization, mass production, and automation) but includes new technologies such as the Internet of Things (IoT)[2], artificial intelligence (AI), Cloud Infrastructure, machine learning, robotics, and additive manufacturing. AI is a key component of Industry 4.0. It refers to the ability of machines to perform tasks that typically require human intelligence, such as learning, problem solving, and decision making. AI[3] is a key component of Industry 4.0. AI is being used in a wide range of applications in Industry 4.0, including predictive maintenance, autonomous robots, and intelligent supply chains. The integration of AI with other technologies is expected to lead to even greater efficiency, productivity, and innovation in manufacturing and other industries. Overall, Industry 4.0 and AI are expected to significantly impact many aspects of society, from the economy to healthcare to transportation, like the way the previous industrial revolutions did.

15.2 FACTORS CONTRIBUTING TOWARDS THE ECONOMIC DIVIDE BETWEEN DEVELOPED AND DEVELOPING NATIONS

The economic divide between developed and developing nations is a complex issue that is shaped by various factors, including historical, political, social, and economic factors. Below are some of the key factors that contribute to this divide:

Historical factors: Developed nations have had a head start in economic development, having gone through industrial revolutions and established advanced economies many decades ago. In contrast, many developing nations have only recently gained independence from colonial powers and have struggled to establish stable governments and economies.[4]

Political factors: Political instability, corruption, and weak institutions can undermine economic growth and development in developing nations. Developed nations tend to have more stable political systems and stronger institutions, which can facilitate economic growth.[5]

Social factors: Social factors such as education, health, and gender equality can impact economic development. Developed nations tend to have higher levels of education and better health outcomes, which can lead to a more productive workforce. The workers in developed and developing nations are being substituted by AI. In addition, developed nations tend to have more gender equality, which can help promote economic growth by expanding the pool of human capital.[6]

Economic factors: Economic factors such as access to capital, technology, and markets can also impact economic development. Developed nations tend to have more advanced financial systems, better technology, and more diversified economies, which can help them attract investment and promote growth.[7]

The post COVID-19 era is full of many kinds of short-term and long-term risks across the globe, and technological risks will create more inequalities. The economic divide is not stated in the World Economic Forum's Global Risk Report, but it needs to gain equal attention. While there are no easy solutions to bridging this divide, efforts to promote economic growth and development in developing nations can include investing in education and health, promoting good governance and strong institutions, and improving access to capital and technology.

15.3 HISTORICAL BACKGROUND OF THE PREVIOUS INDUSTRIAL REVOLUTIONS

The Industrial Revolution refers to a series of significant technological and socio-economic changes that took place in Europe and North America during the 18th and 19th centuries. There were three major industrial revolutions, each marked by different technological innovations and transformations in the economic system.

- The First Industrial Revolution (1760–1840) began in Great Britain, where new inventions and innovations dramatically improved manufacturing processes, leading to increased productivity and profits. The revolution was fueled by new energy sources, such as steam power and the use of machines in factories. The most significant inventions of this era included the steam engine, spinning jenny, power loom, and cotton gin, which revolutionized the textile industry. The growth of transportation systems, such as the development of railroads and steamships, also played a significant role in the expansion of international trade.[8]

- The Second Industrial Revolution (1870–1914) was marked by the introduction of new forms of power, such as electricity, gas, and oil. The most significant innovations of this era included the telephone, typewriter, and assembly line, which enabled the mass production of goods. The Second Industrial Revolution also saw significant advancements in transportation, including the development of the automobile, which transformed personal transportation and the growth of cities.[9]
- The Third Industrial Revolution (1969–2007) is also known as the Digital Revolution. It was marked by the development of new technologies such as computers, the internet, and telecommunications. These innovations transformed communication, information exchange, and the way people work and interact. The Third Industrial Revolution also brought about significant changes in the global economy, including the growth of international trade and the rise of global corporations.[10]

The impact of the Industrial Revolution was far-reaching and shaped modern societies in many ways. It led to significant improvements in living standards, increased urbanization, and the growth of the middle class. However, it also brought about significant social and economic inequalities, including the exploitation of workers and the environmental degradation caused by industrialization.

The Industrial Revolution was a period of significant technological and socioeconomic changes that had a profound impact on modern societies. The three industrial revolutions were marked by different technological innovations and transformations in economic systems, each leading to new forms of power, production, and communication.

15.4   IMPACTS OF INDUSTRIAL REVOLUTION 4.0 AND AI ON THE ECONOMY

The Fourth Industrial Revolution, which is characterized by the integration of advanced technologies such as AI, robotics, and the Internet of Things (IoT), is having a profound impact on the global economy. Here are some key ways in which Industrial Revolution 4.0 and AI are impacting the economy:

1. **Increased productivity:** The integration of advanced technologies such as AI and robotics is leading to increased productivity across various industries. By automating repetitive tasks and improving production efficiency, businesses are able to produce more goods and services in less time, leading to increased output.[11]

2. **Job displacement and creation:** The rise of AI and automation is leading to job displacement in some industries, as machines are able to perform tasks that were previously done by humans. However, new jobs are also being created in industries related to AI and robotics, such as data analysis and software engineering.

3. **Enhanced customer experience:** AI is being used to improve the customer experience by enabling businesses to provide personalized recommendations and customer service. Chatbots, for example, can provide 24/7 customer support and respond to customer inquiries in real time.

4. **Improved decision making:** AI is being used to improve decision making in various industries, such as finance and healthcare. By analyzing large amounts of data, AI can help businesses identify trends and make more informed decisions.

5. **Disruption of traditional business models:** The Fourth Industrial Revolution is disrupting traditional business models by enabling new entrants to enter industries and compete with established businesses. For example, the rise of e-commerce has disrupted traditional brick-and-mortar retail, while ride-sharing services have disrupted traditional taxi services.

6. **Increased economic growth:** Despite concerns about job displacement and other challenges, the integration of AI and advanced technologies is expected to lead to increased economic growth in the long term. A report by PwC estimated that AI could contribute up to $15.7 trillion to the global economy by 2030.[12]

Overall, the Fourth Industrial Revolution and the rise of AI are transforming the global economy in a variety of ways. While there are challenges and risks associated with these developments, they also offer significant opportunities for businesses and economies that are able to adapt and take advantage of them.[13]

There have been several studies examining the potential effects of AI and the Fourth Industrial Revolution on the global economic divide. Here are some key findings:

1. **Automation could lead to job displacement and income inequality:** According to a report by the World Economic Forum, the rise of AI and automation could lead to the displacement of 75 million jobs by 2022. While new jobs may also be created, there is a risk that the jobs lost will be low-skilled jobs that pay lower wages, leading to greater income inequality.[14]

2. **Education and re-skilling will be critical:** To address the potential job displacement caused by automation, education and re-skilling will be critical. Workers will need to acquire new skills to stay relevant in the job market. Governments, businesses, and educational institutions will need to work together to provide training and education opportunities for workers.[15]

3. **AI could lead to increased productivity and economic growth:** While there are concerns about job displacement and income inequality, AI and automation could also lead to increased productivity and economic growth. A report by PwC estimated that AI could contribute up to $15.7 trillion to the global economy by 2030.

4. **There could be a widening skills gap:** As new technologies and skills become more important in the job market, there is a risk of a widening skills gap between developed and developing countries. Developed countries with advanced education systems and access to technology may have an advantage over developing countries in the race to adopt AI and automation.

5. **There is potential for AI to benefit developing countries:** Despite the risks of a widening economic divide, there is also potential for AI to benefit developing countries. For example, AI could improve access to healthcare and education in low-income areas. It could also help small businesses in developing countries to compete in the global marketplace.[16]

Overall, the potential effects of AI and the Fourth Industrial Revolution on the global economic divide are complex and multifaceted. While there are risks of job displacement and income inequality, there is also potential for increased productivity and economic growth. Education and re-skilling will be critical to help workers stay relevant in the job market, and there is a need for cooperation between governments, businesses, and educational institutions to address the potential challenges and opportunities of AI and automation.

15.5   POSITIVE EFFECTS OF AI AND INDUSTRIAL REVOLUTION 4.0 ON BRIDGING THE ECONOMIC DIVIDE

The Fourth Industrial Revolution (Industry 4.0) and the rise of artificial intelligence (AI) have the potential to reduce the economic divide by improving economic productivity and creating new opportunities for growth. Some of the positive effects of AI and Industry 4.0 on the economic divide include:

1. **Increased productivity and efficiency:** The use of AI and Industrial Revolution 4.0 technologies can significantly increase productivity and efficiency in various industries. This leads to a reduction in production costs and an increase in profitability. The increased productivity and efficiency can contribute to the growth of the economy and provide more job opportunities.[17]

2. **Reduced costs and improved quality of products:** AI and Industrial Revolution 4.0 technologies can help businesses reduce costs by automating tasks and optimizing processes. This can lead to improved product quality and consistency, which can ultimately result in increased customer satisfaction and loyalty.[18]

3. **Improved access to technology in developing countries:** AI and Industrial Revolution 4.0 technologies can be used to bridge the technological divide between developed and developing countries. By providing access to technology, businesses in developing countries can improve their productivity and competitiveness, which can lead to economic growth.

4. **Creation of new job opportunities:** AI and Industrial Revolution 4.0 technologies are expected to create new job opportunities in various

fields such as software development, robotics, data analysis, and cy-bersecurity. The demand for skilled professionals in these areas is ex-pected to increase, leading to a reduction in unemployment rates and an increase in job security.[19]

5. **Better decision making:** AI algorithms can analyze large amounts of data and provide insights that can improve decision making in various indus-tries. This can lead to better investment decisions, product development, and resource allocation. This can also lead to better outcomes for con-sumers, such as lower prices and better quality products, which can benefit low-income individuals and reduce economic inequality.

Overall, AI and Industrial Revolution 4.0 technologies have the potential to posi-tively impact economic growth and reduce the economic divide between developed and developing countries. However, it is essential to ensure that the benefits of these technologies are distributed equitably to all members of society.

15.6 NEGATIVE EFFECTS OF AI AND INDUSTRIAL REVOLUTION 4.0 ON WIDENING THE ECONOMIC DIVIDE

The negative effects of AI and the Fourth Industrial Revolution on widening the economic divide are as follows:

1. **Loss of jobs due to automation:** As companies adopt AI and Industrial Revolution 4.0 technologies, they may automate tasks that were previously performed by humans. This can result in job displacement, particularly in industries that rely heavily on manual labor. Workers who lack skills in the use of technology may find it challenging to secure new jobs, leading to increased unemployment and income inequality.

2. **Growing income inequality within countries:** The adoption of AI and Industrial Revolution 4.0 technologies may widen the income gap between those who possess skills in the use of these technologies and those who do not. Highly skilled workers may earn significantly higher wages than low-skilled workers, leading to increased income inequality.[20]

3. **The concentration of wealth and power among a few large technology corporations:** As companies adopt AI and Industrial Revolution 4.0 technologies, those who are successful in these industries may accumulate significant wealth and power. This can lead to the concentration of wealth and power among a few large technology corporations, exacerbating the economic divide.

4. **Increased economic dependence on technological advancement in developing countries:** Developing countries may be at a disadvantage in the adoption of AI and Industrial Revolution 4.0 technologies, as they may lack the necessary infrastructure, resources, and skills. This can lead to increased economic dependence on developed countries and multinational corporations, further widening the economic divide.

5. **The digital divide:** The adoption of AI and Industrial Revolution 4.0 technologies is highly dependent on access to digital infrastructure. In developing countries or marginalized communities, lack of access to technology can exacerbate the digital divide and further widen the economic divide.[21]

6. **Privacy and security concerns:** The use of AI and Industrial Revolution 4.0 technologies can raise privacy and security concerns. With increasing reliance on data collection and processing, there is a risk that sensitive information may be mishandled or exploited, leading to harm to individuals and society.[22]

In conclusion, while AI and Industrial Revolution 4.0 technologies have the potential to positively impact economic growth, they can also exacerbate the economic divide. It is important to address these negative effects to ensure that the benefits of these technologies are distributed equitably and do not leave behind vulnerable members of society.[23]

15.7   COUNTRIES WHERE AI AND INDUSTRIAL REVOLUTION 4.0 HAVE BRIDGED THE ECONOMIC DIVIDE

There are several countries where AI and Industrial Revolution 4.0 have played a significant role in bridging the economic divide. Here are a few examples:

1. **China:** China has made significant strides in implementing AI and Industry 4.0 technologies, and it has become a global leader in this field. The Chinese government has invested heavily in the development of AI and other emerging technologies. As a result, China has seen rapid economic growth, and the technology sector has become a key driver of the country's economy.

2. **United States:** The United States is also a leader in the development and implementation of AI and Industry 4.0 technologies. Companies like Amazon, Google, and Microsoft have been at the forefront of this revolution, and they have played a significant role in driving economic growth in the country.[24]

3. **Germany:** Germany is known for its advanced manufacturing capabilities, and it has been quick to adopt Industry 4.0 technologies. The German government has invested heavily in research and development, and the country's manufacturing sector has seen significant gains as a result.

4. **South Korea:** South Korea is another country that has embraced AI and Industry 4.0 technologies. The government has invested heavily in these areas, and the country has become a leader in the development of autonomous vehicles, robotics, and other emerging technologies.[25]

5. **Singapore:** Singapore has also been quick to adopt AI and Industry 4.0 technologies. The government has invested heavily in research and development, and the country has become a hub for tech startups and innovative companies. As a result, Singapore has seen rapid economic growth and has become one of the most prosperous countries in the world.[26]

15.8　COUNTRIES WHERE AI AND INDUSTRIAL REVOLUTION 4.0 HAVE WIDENED THE ECONOMIC DIVIDE

There are several examples of countries where AI and the Fourth Industrial Revolution have widened the economic divide between different groups of people. Here are a few examples:

1. **India:** India is a country with a rapidly growing economy, but the benefits of this growth have not been equally shared. The country has a large population of poor people who have not benefited from the new technologies and industries that have emerged in recent years.[27]
2. **Brazil:** Brazil is another country where the economic divide has widened as a result of AI and the Fourth Industrial Revolution. The country has a large population of poor people who have not been able to access the new technologies and industries that have emerged in recent years.
3. **Russia:** Russia is a country that has made significant investments in AI and other Fourth Industrial Revolution technologies, but these advancements have not led to widespread economic growth. Instead, they have mostly benefited a small group of wealthy elites, while the majority of the population continues to struggle with poverty and unemployment.
4. **South Africa:** South Africa is a country that has struggled with high levels of income inequality for decades, and the Fourth Industrial Revolution has not made things better. Instead, it has led to the creation of new industries and jobs that are only accessible to a small group of highly skilled and educated people, leaving many others behind.[28]
5. **Mexico:** Mexico is a country that has embraced AI and Fourth Industrial Revolution technologies in recent years, but these advancements have not led to widespread economic growth or reduced inequality. Instead, they have mostly benefited a small group of wealthy elites, while many others struggle to find work and make ends meet.

15.9　POTENTIAL LONG-TERM EFFECTS OF AI AND INDUSTRIAL REVOLUTION 4.0 ON THE ECONOMY AND THE GLOBAL ECONOMIC DIVIDE

The potential long-term effects of AI and the Fourth Industrial Revolution on the economy and the global economic divide are complex and multifaceted. Here are some potential scenarios:

1. **Widening economic divide:** One possible long-term effect of AI and the Fourth Industrial Revolution is that they could widen the economic divide between the rich and poor. As automation and AI replace human labour in many industries, those with the skills and education needed to work with these technologies may benefit, while those without those skills may be left behind.
2. **Increased productivity:** On the other hand, AI and the Fourth Industrial Revolution have the potential to increase productivity and output, leading to

economic growth and increased prosperity. If the benefits of this growth are shared equally, it could lead to a reduction in global economic inequality.

3. **New job opportunities:** While automation and AI may displace some workers, they could also create new job opportunities in emerging industries such as data analysis, software engineering, and robotics.

4. **Technological advancements:** The Fourth Industrial Revolution has the potential to drive technological advancements that could transform the global economy, leading to new industries and opportunities for growth. However, this may also lead to increased competition and disruption for existing industries.

5. **Digital divide:** The digital divide, or the gap between those who have access to technology and the internet and those who do not, could widen as AI and the Fourth Industrial Revolution become more prevalent. This could further exacerbate existing economic disparities.[29]

Overall, the long-term effects of AI and the Fourth Industrial Revolution on the global economy and economic divide will depend on how these technologies are implemented and regulated. Policy makers and industry leaders must work to ensure that the benefits of these technologies are shared equitably, and that they are used to create a more just and prosperous global economy. This will require investments in education and workforce development, as well as measures to address the digital divide and ensure that everyone has access to the opportunities created by these technologies.[30]

15.10 CONCLUSION

The Fourth Industrial Revolution and the advancements in AI have the potential to either bridge or widen the economic divide between developed and developing nations. The historical background of previous industrial revolutions shows that technological advancements have had both positive and negative impacts on the economy.

The First Industrial Revolution, which began in the late 18th century, saw the transition from hand production methods to machines, leading to increased productivity and economic growth. However, it also caused a shift in the distribution of wealth, leading to increased inequality and poverty.

The Second Industrial Revolution saw the rise of mass production and electrification, further boosting economic growth and leading to the creation of new jobs. However, it also caused social and environmental issues.

The Third Industrial Revolution, also known as the Digital Revolution, saw the advent of computers, the internet, and automation. While it created new jobs and increased productivity, it also led to job displacement and income inequality.

The Fourth Industrial Revolution and AI are expected to bring significant changes to the economy, including increased automation, robotics, and the Internet of Things. While this could lead to increased productivity and efficiency, it could also lead to job displacement, particularly in developing nations. However, the adoption of these technologies could also create new jobs and industries, leading to economic growth and bridging the economic divide.

It is crucial to note that the effects of Industry 4.0 and AI on the economy are complex and varied. Their long-term impact will depend on factors such as government policies, investment in education and training, and the ability of businesses to adapt and adopt new technologies. Therefore, it is important to approach these changes with caution, considering their potential impact on various segments of society, particularly those in developing nations.

Overall, the impact of AI and Industrial Revolution 4.0 on the economic divide will depend on how well governments and societies can manage the technology's development and distribution. Countries that are proactive in implementing policies and strategies to ensure that the benefits of AI and Industrial Revolution 4.0 are shared equitably are likely to bridge the economic divide. On the other hand, countries that lack the necessary infrastructure, skills, and resources may fall further behind.

15.11   IMPLICATIONS OF THE RESEARCH

The implications of the historical background of previous industrial revolutions and the potential impact of Industry 4.0 and AI on the economy are significant. It suggests that the adoption of new technologies could lead to both positive and negative consequences.

One of the key implications of this research is that the historical background of previous industrial revolutions can inform our understanding of the potential effects of AI and Industrial Revolution 4.0 on the global economy. While technological progress has the potential to drive economic growth, it can also create new forms of inequality. Therefore, policy makers and stakeholders must take a careful approach to ensure that the benefits of these technologies are distributed equitably.

Another implication of this research is the importance of proactive government policies and strategies to ensure that the benefits of AI and Industrial Revolution 4.0 are shared equitably. Countries that lack the necessary infrastructure, skills, and resources may fall further behind if they do not take action to implement policies that will enable them to benefit from these technologies. Conversely, countries that are proactive in implementing such policies and strategies are more likely to bridge the economic divide.

Furthermore, this research highlights the need for ongoing monitoring and evaluation of the impact of AI and Industrial Revolution 4.0 on the global economy and economic divide. The long-term effects of these technologies are yet to be seen, and ongoing research will be necessary to understand how they will shape the future of the global economy.

On the positive side, Industry 4.0 and AI could lead to increased productivity and efficiency, leading to economic growth and the creation of new industries and jobs. This could help to bridge the economic divide by providing opportunities for individuals and businesses in developing nations to participate in the global economy.

On the negative side, the adoption of new technologies could lead to job displacement, particularly in developing nations where labour costs are relatively low. This could widen the economic divide by exacerbating income inequality and reducing economic opportunities for individuals in these countries.

Therefore, it is essential to consider the potential implications of Industry 4.0 and AI on the economy carefully. Governments and businesses should work together to ensure that the adoption of new technologies is done in a way that benefits everyone, particularly those in developing nations. This could include investing in education and training to prepare individuals for new jobs and industries, creating policies to support job creation and economic growth, and ensuring that the benefits of technological advancements are distributed fairly across society.

The study emphasizes the importance of a careful and proactive approach to the implementation of AI and Industrial Revolution 4.0 technologies to ensure that they have a positive impact on the economic divide between developed and developing nations. With the right policies and strategies in place, these technologies have the potential to drive economic growth and innovation while reducing inequality and while the potential impact of Industry 4.0 and AI on the economy is significant, their adoption must be approached with caution. By taking a thoughtful and inclusive approach, governments and businesses can help to ensure that these new technologies benefit everyone, bridging the economic divide and promoting long-term economic growth and prosperity.

15.12 SUGGESTIONS FOR FUTURE RESEARCH IN THIS AREA

Following are the suggestions for future research in this area:

- **Long-term effects:** While there have been several studies on the potential effects of AI and Industrial Revolution 4.0 on the global economic divide, it is essential to continue monitoring the long-term effects of these technologies. Future research could explore the ongoing impact of AI and Industrial Revolution 4.0 on economic growth, innovation, and inequality in both developed and developing nations.
- **Regional and national analysis:** It is also important to conduct more in-depth regional and national analyses of the impact of AI and Industrial Revolution 4.0 on the economic divide. This could help policy makers identify specific areas of concern and implement targeted policies and strategies to bridge the gap between developed and developing nations.
- **Equity and social impact:** Future research could explore the equity and social impact of AI and Industrial Revolution 4.0 on different groups within society, such as marginalized communities and women. Such research could help identify the potential benefits and challenges of these technologies for different groups and inform policies and strategies to ensure that the benefits of AI and Industrial Revolution 4.0 are distributed equitably.
- **Governance and regulation:** Future research could explore the governance and regulation of AI and Industrial Revolution 4.0 technologies, particularly in developing nations. Such research could help identify best practices and guide the development of policies and regulations that promote responsible innovation and minimize potential negative impacts.
- **Skills and education:** Future research could explore the role of skills and education in bridging the economic divide. Specifically, it could investigate

how to develop and deliver effective training and education programs that enable individuals in developing nations to acquire the skills necessary to participate in the digital economy.

- **Cross-disciplinary research:** Future research could explore the intersection of AI and Industrial Revolution 4.0 with other disciplines, such as sociology, economics, and political science. Specifically, it could investigate how social, economic, and political factors influence the impact of AI and Industrial Revolution 4.0 on the economic divide, and how these technologies, in turn, affect these factors.
- **Environmental impact:** Future research could examine the environmental impact of AI and Industrial Revolution 4.0 technologies, particularly in developing nations. Specifically, it could investigate how to minimize the negative environmental impact of these technologies and promote sustainable development in developing nations.

In summary, future research in this area should continue to explore the potential impacts of AI and the Fourth Industrial Revolution on the economic divide between developed and developing nations, with a focus on long-term effects, equity and social impact, regional and national analysis, and governance and regulation, skills and education, cross-disciplinary research, and environmental impact. By addressing these topics, future research could provide valuable insights into how AI and Industrial Revolution 4.0 can be used to bridge the economic divide between developed and developing nations and also such research could provide valuable insights to inform policy making and guide the responsible development and deployment of these technologies.

NOTES

1 Lee, Min Goo, Jin Han Yun, Andreas Pyka, Dong Il Won, Fumio Kodama, Giovanni Schiuma, Hang A Park, et al. "How to Respond to the Fourth Industrial Revolution, or the Second Information Technology Revolution? Dynamic New Combinations between Technology, Market, and Society through Open Innovation." Journal of Open Innovation. Springer Science+Business Media, June 21, 2018. https://doi.org/10.3390/joitmc4030021.
2 Madakam, Somayya, R. Ramaswamy, and Siddharth K. Tripathi. "Internet of Things (IoT): A Literature Review." Journal of Computer and Communications 03, no. 05 (May 25, 2015): 164–73. https://doi.org/10.4236/jcc.2015.35021
3 Hancock, Jeffrey T., Mor Naaman, and Karen Levy. "AI-Mediated Communication: Definition, Research Agenda, and Ethical Considerations." Journal of Computer-Mediated Communication 25, no. 1 (March 23, 2020): 89–100. https://doi.org/10.1093/jcmc/zmz022.
4 Arocena, R., & Sutz, J. (2002). Innovation systems and developing countries. DRUID (Danish Research Unit for Industrial Dynamics) Working Paper, 2(05).
5 Rothstein, B. (2011). The quality of government: Corruption, social trust, and inequality in international perspective. University of Chicago Press.
6 Rothstein, B., & Uslaner, E. M. (2005). All for all: Equality, corruption, and social trust. World politics, 58(1), 41–72. Retrieved from: https://www.cambridge.org/core/iournals/world-politics/article/all-for-all-equalitv-corruption-and-social-trust/09B64F404EB0F753E78680B70A9ABEDB

7   Rothstein, B., & Uslaner, E. M. (2005). All for all: Equality, corruption, and social trust. World politics, 58(1), 41-72. https://www.cambridge.org/core/iournals/world-politics/article/all-for-all-equalitv-corruption-and-social-trust/09B64F404EB0F753E78680B70A9ABEDB

8   Freeman, Chris, and Francisco Louçã. As Time Goes By: From the Industrial Revolutions to the Information Revolution. OUP Oxford, 2001.

9   Mokyr, Joel and Robert H. Strotz. "The Second Industrial Revolution, 1870-1914." (2000).

10 Josef Taalbi, Origins and pathways of innovation in the third industrial revolution, *Industrial and Corporate Change*, Volume 28, Issue 5, October 2019, Pages 1125–1148, https://doi.org/10.1093/icc/dty053

11 Park, SC. The Fourth Industrial Revolution and implications for innovative cluster policies. AI & Soc 33, 433–445 (2018). https://doi.org/10.1007/s00146-017-0777-5

12 PWC website: https://www.pwc.com/gx/en/issues/analytics/assets/pwc-ai-analysis-sizing-the-prize-report.pdf

13 Xu, Min, Jeanne M. David, and Suk Kim. "The Fourth Industrial Revolution: Opportunities and Challenges." International Journal of Financial Research 9, no. 2 (February 5, 2018): 90. https://doi.org/10.5430/ijfr.v9n2p90.

14 Loris Caruso, "Digital Innovation and the Fourth Industrial Revolution: Epochal Social Changes?" AI & Society 33, no. 3 (August 1, 2018): 379–92, https://doi.org/10.1007/s00146-017-0736-1.

15 Zervoudi, Evanthia K. "Fourth Industrial Revolution: Opportunities, Challenges, and Proposed Policies." IntechOpen EBooks, January 21, 2020. https://doi.org/10.5772/intechopen.90412.

16 McKinsey & Company. "Notes from the AI Frontier: Modeling the Impact of AI on the World Economy," September 4, 2018. https://www.mckinsey.com/featured-insights/artificial-intelligence/notes-from-the-ai-frontier-modeling-the-impact-of-ai-on-the-world-economy.

17 Aly, H. (2022), "Digital transformation, development and productivity in developing countries: is artificial intelligence a curse or a blessing?", Review of Economics and Political Science, Vol. 7 No. 4, pp. 238-256. https://doi.org/10.1108/REPS-11-2019-0145

18 Ghoreishi, Malahat, and Ari Happonen. "Key Enablers for Deploying Artificial Intelligence for Circular Economy Embracing Sustainable Product Design: Three Case Studies." Nucleation and Atmospheric Aerosols, May 5, 2020. https://doi.org/10.1063/5.0001339.

19 Hoosain, Mohamed Sameer, Babu Sena Paul, and Seeram Ramakrishna. 2020. "The Impact of 4IR Digital Technologies and Circular Thinking on the United Nations Sustainable Development Goals" Sustainability 12, no. 23: 10143. https://doi.org/10.3390/su122310143

20 Hoosain, Mohamed Sameer, Babu Sena Paul, and Seeram Ramakrishna. 2020. "The Impact of 4IR Digital Technologies and Circular Thinking on the United Nations Sustainable Development Goals" Sustainability 12, no. 23: 10143. https://doi.org/10.3390/su122310143

21 Cowie, Paul, Leanne Townsend, and Koen Salemink. "Smart Rural Futures: Will Rural Areas Be Left behind in the 4th Industrial Revolution?" Journal of Rural Studies. October 1, 2020. https://doi.org/10.1016/j.jrurstud.2020.08.042.

22 Lee, DonHee, and Seong No Yoon. 2021. "Application of Artificial Intelligence-Based Technologies in the Healthcare Industry: Opportunities and Challenges" International Journal of Environmental Research and Public Health 18, no. 1: 271. https://doi.org/10.3390/ijerph18010271

23 Aly, H. (2022), "Digital transformation, development and productivity in developing countries: is artificial intelligence a curse or a blessing?", Review of Economics and Political Science, Vol. 7 No. 4, pp. 238-256. https://doi.org/10.1108/REPS-11-2019-0145

24 Morrar, Rabeh & Arman, Husam. (2017). The Fourth Industrial Revolution (Industry 4.0): A Social Innovation Perspective. Technology Innovation Management Review. 7. 12-20. 10.22215/timreview/1117.

25 Lee, Keun, Chan-Yuan Wong, Patarapong Intarakumnerd, and Chaiyatorn Limapornvanich. "Is the Fourth Industrial Revolution a Window of Opportunity for Upgrading or Reinforcing the Middle-Income Trap? Asian Model of Development in Southeast Asia." Journal of Economic Policy Reform 23, no. 4 (October 1, 2020): 408–25. https://doi.org/10.1080/17487870.2019.1565411.

26 Liao, Y., Loures, E. R., Deschamps, F., Brezinski, G., & Venâncio, A. (2017). The impact of the fourth industrial revolution: a cross-country/region comparison. Production, 28, e20180061. DOI: 10.1590/0103-6513.20180061

27 Mhlanga, David. "Artificial Intelligence in the Industry 4.0, and Its Impact on Poverty, Innovation, Infrastructure Development, and the Sustainable Development Goals: Lessons from Emerging Economies?" *Sustainability* 13, no. 11 (May 21, 2021): 5788. https://doi.org/10.3390/su13115788.

28 Manda, More Ickson, and Soumaya Ben Dhaou. "Responding to the Challenges and Opportunities in the 4th Industrial Revolution in Developing Countries." International Conference on Theory and Practice of Electronic Governance, April 3, 2019. https://doi.org/10.1145/3326365.3326398.

29 Lee, MinHwa, JinHyo Joseph Yun, Andreas Pyka, DongKyu Won, Fumio Kodama, Giovanni Schiuma, HangSik Park, Jeonghwan Jeon, KyungBae Park, KwangHo Jung, Min-Ren Yan, SamYoul Lee, and Xiaofei Zhao. 2018. "How to Respond to the Fourth Industrial Revolution, or the Second Information Technology Revolution? Dynamic New Combinations between Technology, Market, and Society through Open Innovation" Journal of Open Innovation: Technology, Market, and Complexity 4, no. 3: 21. https://doi.org/10.3390/joitmc4030021

30 Caruso, L. Digital innovation and the fourth industrial revolution: epochal social changes? AI & Soc 33, 379–392 (2018). https://doi.org/10.1007/s00146-017-0736-1

REFERENCES

Bughin, J., Seong, J., Manyika, J., Chui, M., & Joshi, R. (2018). *Notes from the AI frontier: Modeling the impact of AI on the world economy.* McKinsey Global Institute, 4.

Caruso, L. (2018). Digital innovation and the fourth industrial revolution: epochal social changes? *AI & Society*, 33(3), 379–392.

Cowie, P., Townsend, L., & Salemink, K. (2020). Smart rural futures: Will rural areas be left behind in the 4th industrial revolution?. *Journal of Rural Studies*, 79, 169–176.

Ghoreishi, M., & Happonen, A. (2020, May). Key enablers for deploying artificial intelligence for circular economy embracing sustainable product design: Three case studies. In AIP conference proceedings (Vol. 2233, No. 1, p. 050008). AIP Publishing LLC.

Hoosain, M. S., Paul, B. S., & Ramakrishna, S. (2020). The impact of 4IR digital technologies and circular thinking on the United Nations sustainable development goals. *Sustainability*, 12(23), 10143.

Lee, J., Davari, H., Singh, J., & Pandhare, V. (2018). Industrial Artificial Intelligence for industry 4.0-based manufacturing systems. *Manufacturing Letters*, 18, 20–23.

Lee, D., & Yoon, S. N. (2021). Application of artificial intelligence-based technologies in the healthcare industry: Opportunities and challenges. *International Journal of Environmental Research and Public Health*, 18(1), 271.

Madakam, S., Lake, V., Lake, V., & Lake, V. (2015). Internet of Things (IoT A literature review. *Journal of Computer and Communications*, 3(05), 164).

Morrar, R., Arman, H., & Mousa, S. (2017). The fourth industrial revolution (Industry 4.0): A social innovation perspective. *Technology Innovation Management Review*, 7(11), 12–20.

16 Digitalization

Developments and Disputations

Pinki Chugh, Harsh Verma, and Isha Ali
UIT, Uttaranchal University, Dehradun

16.1 INTRODUCTION

Digitalization is an area that is quickly developing and has the potential to revolutionize several industries as well as change how people live and work. Artificial intelligence (AI) and digital technologies are being integrated into a variety of industries, which is resulting in increased automation, better decision making, and more precise data analysis. This study intends to investigate the state of digitalization at the moment, its consequences on businesses and society, and potential future developments.

The ability to swiftly and effectively collect and analyze massive amounts of data is one of the main advantages of digitalization. Digitalization can be used to optimize routes, enhance traffic flow, and lower emissions in the transportation sector, which is another significant use. This may result in fewer negative environmental effects and more effective transportation networks. Additionally, digitalization is utilized in manufacturing to automate monotonous operations, increase productivity, and cut costs. The productivity and competitiveness of the company in the international market also increases as a result.

The widespread use of digitalization is hampered by several issues and restrictions, though. The possible loss of jobs due to increased automation is one of the key worries. The use of AI raises further ethical questions, such as the possibility of prejudice and the necessity of accountability and openness. Aside from these issues, there are worries about how AI-powered systems might be mishandled and how they would affect privacy and security.

16.2 DIGITALIZATION – THE CURRENT SCENARIO

The current state of digitalization is characterized by a rapid integration of AI and digital technology across multiple sectors. The number of businesses investing in AI and machine learning has significantly increased in recent years, with which the worldwide AI market is expected to reach $190 billion by 2025. This growth is driven by the many potential benefits of AI and digital technology, including improved efficiency, cost savings, and enhanced decision making.

Digitalization has the potential to significantly raise production and efficiency across a variety of industries. As with any new technology, it's crucial to thoroughly weigh the advantages and disadvantages as well as any relevant ethical issues. To make sure that AI serves the interests of society as a whole, it is our duty as researchers and practitioners in the area to create and deploy it responsibly and deliberately.

Due to digitalization and AI, the **software market** has been expanding at breakneck speed; the latest AI market forecast, for example, shows that the industry is driven by the uptick in the category's use cases. However, growth is not confined to the software industry, as AI is also expected to leave a positive economic footprint all around. The following are key AI market statistics that are worth knowing (Figure 16.1).

The AI software market's global annual revenue is currently over $50 billion.

As per the reports, the annual revenue in 2018 was over $10.1B, 2019 was approx. $14.69B, 2020 was $22.59B, 2021 was $34.87B, and 2022 was $51.27B. The current revenue is over $50B, and for the future, the expected revenue is $70.94B in 2023, $94.41B in 2024, and $126B in 2025.

AI-based solutions, for instance, have the power to completely change how diseases are identified and treated in the **healthcare industry**. AI-powered diagnostic technologies, like deep learning algorithms, may examine medical imaging like CT scans and MRIs and find early disease indications, improving the likelihood of a successful recovery. The creation of new therapies and medications is also being sped up by AI-based research techniques like natural language processing, which analyze vast amounts of data. This results in the identification of new pharmacological targets, methods for estimating drug effectiveness, and methods for detecting drug adverse effects, all of which result in the potential saving of lives.

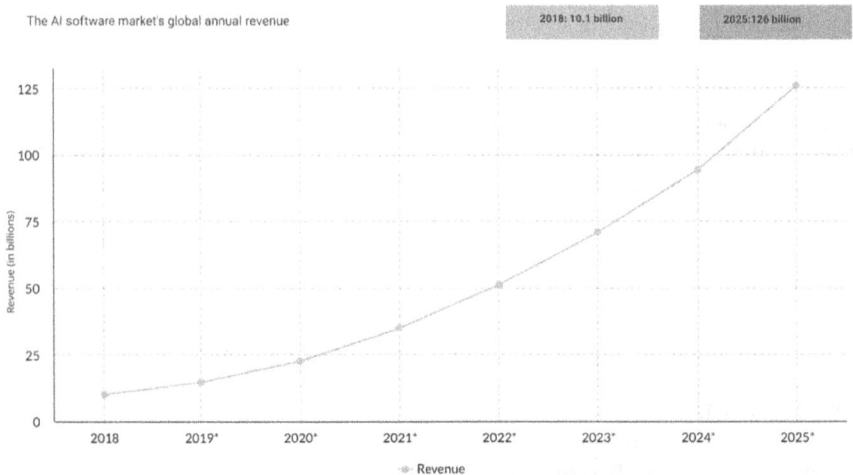

FIGURE 16.1   The AI software market's global annual revenue (2018 to 2025).

Another industry that has been heavily impacted by digitalization is **finance**. Digitalization has been used to detect fraudulent transactions, manage risk, and optimize investment decisions. For example, AI algorithms are used to identify patterns in financial data that may indicate fraudulent activity. Additionally, AI is used to analyze large amounts of data from financial markets to identify trends and make more informed investment decisions.

The **transportation sector** has benefited significantly from digitalization as well. Artificial intelligence-powered autonomous vehicles and drones eliminate accidents brought on by human error, boost productivity, and enhance safety. AI-powered logistics and routing systems also optimize travel routes, cutting costs and emissions. For instance, self-driving vehicles eliminate the need for human drivers, saving logistics organizations money and lowering transportation-related emissions. Drones are also used for distant infrastructure inspection, such as checking electricity lines, which are dangerous for human operators.

Significant advantages of AI-based technology have also been realized in the **banking sector**. While chatbots and customer care systems powered by AI can improve customer experience, risk management systems powered by AI can identify and stop fraud. For instance, AI-based systems can identify suspicious transactions or account activity and flag them for more examination. This assists financial organizations in identifying and stopping fraud, money laundering, and other financial crimes. Additionally, chatbots are utilized to offer 24/7 customer assistance, which can help financial institutions cut expenses and enhance the client experience.

Digitalization has had a huge positive impact on **manufacturing** as well. AI-powered robotics and logistics automate tedious activities, improve quality control, and cut expenses. By monitoring and adjusting the temperature, humidity, and other environmental conditions, AI-based systems, for instance, are utilized to regulate and improve the production process. In addition to lowering waste, this raises the quality of the finished product. Automating repetitive processes like welding, painting, and assembling is another application for robotics, powered by AI. This increases productivity and lowers labour costs.

Digitalization, which automates laborious operations, makes it possible to make more precise decisions, and improves customer experience, has the potential to completely transform several industries.

Manufacturing is going through a digital transformation, and smart technology, data analytics, and connected devices are enabling manufacturers to dramatically increase their efficiency, productivity, and accuracy. Here is the data of integration of digitalization in the manufacturing industry: (Figure 16.2)

Data are actually the new "petrol" that is driving us through the 21st century. Today humans generate 2.5 quintillion bytes of data, which is a humongous figure in itself. If processed, analyzed, and used correctly, these data can help us in every sphere of our lives. Today, we see how **digitalization and education** are related, and how AI is already helping us to learn more efficiently than ever before. After the deadly pandemic of 2019, there has been a steep climb in the screen time of every person out there. Studies show that for a non-heavy user, the screen time rose from 25 hrs a week to 40 hrs a week, and for heavily used youngsters and adults, the screen time spiked from 10 hrs per day to 17.5 hrs per day. While we are still

Integration of Digitalization in Manufacturing industry

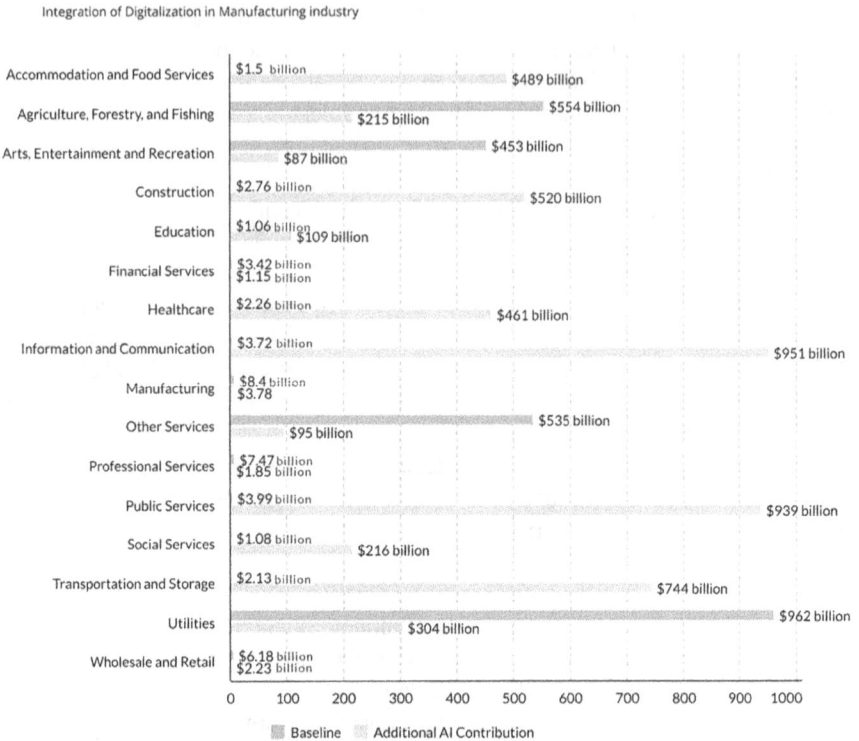

FIGURE 16.2 Integration of digitalization in the manufacturing industry.

debating on ways of reducing screen time for different age groups, using digitalization techniques, we may be able to redirect this issue to a more fruitful solution. While we can never deny the fact that the role of a teacher in children's life is irreplaceable, we have to admit that in the upcoming years, we will see the rise of digitalization in the education sector.

Digitalization is already making its way into the education sphere. Today we can see different EdTech start-ups around, which are providing an enhanced learning experience to every kid. We all know that every kid is unique in himself and that the best learning can be provided to a kid only after analyzing psychological and physiological aspects, individually. Digitalization methods like the use of AI can be used to provide more personalized learning content and experience to every child. The AI models can be trained to analyze various potentials and dimensions of the child's personality to find a more fruitful way of teaching a concept to him so that the kid grasps it easily.

Today, one of the big examples that shows what the face of education may look like shortly is what high-tech schools of Shanghai are experimenting with. The day of a kid in this school starts with wearing a headband, which is interestingly not a usual headband, but an advanced headband loaded with advanced sensors that claim to detect and transfer the neural signals of the brain in real time to a computer. These neutral signals are further processed to detect the concentration levels of the child in real time. The AI dashboards send the data to the computer at the teacher's podium, and then the teacher can pay more attention to the child whose concentration seems to deteriorate while solving a problem. Not just that, but these students also have to wear special uniforms equipped with GPS tracking that helps teachers and parents to find where they are exactly. The surveillance cameras employ computer vision to detect how many times a child yawns or uses his phone during the lectures, and prepare a real-time report of the child. The report is then shared with the teacher and parents so that they can motivate their children to achieve better grades and pay more attention in their lectures.

Another example is from Japan, where scientists are trying to prepare an AI simulation of day-to-day life, which would let children experiment with their decision-making skills and their learning in a safe environment so that even if they fail repeatedly, AI can help them gradually improve without actually letting them face consequences. These simulations have managed to bag quite fruitful results when tested on these grounds.

Today, we are developing technologies that help students around the globe to connect to the best professors around the world and attend their lectures in the comfort of their homes. This has helped us to spread education in most rural and suburban areas, where children don't get access to education due to the lack of good infrastructure. Today, AI software is helping to transcribe the lectures of the world's most renowned professors in the native language of the children, which helps children to connect and learn from these professors in real time. In the upcoming years, what we can expect out of this use-case of AI, is that it will be providing us with a virtual ecosystem, where the world's best teachers will be able to connect with more and more children and impart their knowledge and findings to them, whereas teachers on a local level will be helping these students with invigilating and monitoring their concentration levels individually, which otherwise would be a very tough job for the professors imparting lectures to such a huge crowd of students.

Google has already introduced project LaMDA in Google IO 2021. It is a new natural language processing (NLP) engine that can incorporate and simulate our day-to-day communication to answer even our weirdest questions, without any dead-ends in the conversation. Not just this, but Siri and other chatbots are on the rise these days. They are helping us to accurately find the answers to our daily facts without actually letting us spend hours finding the answers to our questions. We can expect the search engine to advance even more in upcoming times, giving us much more precise and personalized search results in almost no time. These changes in the fundamental building blocks of education will drift our paths from merely memorizing facts and formulas to actually implementing them to get efficient results for our problems, finally helping us shape a better world that empowers the weaker sections of society and brings equal opportunities to everyone.

Digitalization and physical fitness is another area that intrigues and concerns a lot of people in today's generation. In particular, it has revolutionized the physical fitness of humans through advanced fitness equipment that makes our day-to-day home workout much easier and smarter than ever before. But the main question is how AI in physical fitness has revolutionized the physical fitness industry over time. There are a large number of AI fitness benefits. Some of those are listed below:

A personal trainer that is based on AI – In today's super hectic lives, it is nearly impossible to make time for anything, not even our physical fitness. We are so busy rushing to work or the office that we don't get time to look after our health or fitness, and even if we do, the major problem is finding a good and experienced physical fitness trainer quickly. In situations like these, AI becomes our saviour and helps us out. As a matter of fact, personalized gym or fitness trainers are not that economical for everybody, whereas integrated fitness applications with AI are proper and fantastic approaches for those who want their fitness objectives to be achieved.

Human pose estimation technology – The assessment of human poses is a sophisticated technology based on computer vision.

Talking about **digitalization and human psychology**, it is perfectly apt to state that AI is completely dependent on human psychology. Even the mere origin of AI is based on the psychology of the human brain since it is all about how a machine functions just like a human brain without any human operating it or controlling it. Well, it is quite a well-known fact that the future of the world will rely on AI sooner or later, and we need to remain updated to prepare ourselves for the upcoming drastic technological change shortly. And hence, in a future where AI is going to be universal, human psychology is going to stay as an asset that would help individuals to adapt to the vulnerability of the world and change. As the world turns out to be more innovative with each passing day, along with that the requirement for human-based advising and connection also seems to increase. And this results in both human psychology and AI going hand in hand for the best results possible in the technological advancements of our upcoming years. The AI robots these days, that are already in different corporate sectors or that are a part of our day-to-day lives, tend to be operating well when it comes to working as per human psychology. But as it is said, there is always a scope for improvement when it comes to human inventions, and so it is the same for the working of human psychology and AI together in the technical field.

16.3   DISTRACTIONS OF DIGITALIZATION

The broad adoption of digitalization, however, is not without its difficulties and restrictions, as with any new technology. The possible loss of jobs as automation grows is one of the key worries. The use of AI raises further ethical questions, such as the possibility of prejudice and the requirement for accountability and transparency. For instance, if AI systems are educated on biased data, they may continue to make predictions and conclusions that reflect these biases. It is essential to recognize these difficulties and work to make sure that the advantages of artificial digitalization are attained in a just and responsible manner.

The integration of digital pedagogical tools and disruptive technology in education has been a topic of ongoing discussion, with contrasting opinions on its influence on learning outcomes. On one hand, digital technology provides more interactivity and access to vast resources, but on the other, it has been criticized for its effectiveness in comparison to traditional education methods.

"Studies have shown that students who take notes on laptops during lectures perform worse on tests measuring factual content and conceptual understanding when compared to those who take handwritten notes. This is because typing lectures verbatim on laptops doesn't lead to meaningful engagement and internalization of information, unlike the process of selective note-taking that results in better recall".

"Another issue with digital technology in the classroom is the potential for distractions and disruptions. Online content can replace traditional distractions like note-passing, idle chatter, and even the sound of students typing on their keyboards, hindering students' ability to learn and retain the material effectively" (Hembrooke & Gay, 2003). Additionally, students face difficulty balancing between following the lecture and browsing for course-related material on their devices.

We need to admit that with the rise of digitalization in different daily chores, there will be less room for jobs that require routine decision making, or any routine tasks. The bubble of education is now expanding from mere knowledge transfer to discovering newer ways of implementing knowledge to solve daily life problems of the human race. But let us look at the brighter side of this. While there will be a steep decline in routine jobs, there will be a high demand for jobs that require creativity and precise decision making.

Even though this sounds luxurious, AI has led humans to lose the sense of hard work. Nowadays, AI can do wonders, and we won't even have to lift a finger. But this lack of labour and movement has made our bodies lazy and unable to endure big physical hardships. From doing dishes, to doing laundry, to dusting and substituting staircases with hi-tech sensor-based lifts and escalators, smart ACs, and smart gadgets are so smart that they have turned "Jack" into a "Dumbo". Man is so entrapped in this artificial web of comforts that he has lost his power of the five senses. Those senses, which were used as a model for AI, have stopped working now or, we can say, they have been suppressed by machine learning (ML)!

Due to AI, we have been introduced to a lot of screens. Even though they are quite easily compatible with humans, these screens damage our eyes. Whether doing some work in the office or playing online games, travelling somewhere using Google Maps, or ordering food from Zomato, Swiggy, etc. booking tickets, online shopping, and whatnot … we are stuck to screens 24×7. The excess exposure to screens has led a large number of people into wearing spectacles and contact lenses. And sometimes if the usage of screens goes to extremes, then it can lead to permanent loss of vision.

Touch pads and touch screens make up a huge part of AI-powered appliances, and both of them have radically lowered the actual spirit of writing. We first shifted from fountain pens to ink ones and then to keyboards and touch screens, but now we are entering into an era of speech typing. The art of writing has always revolved around the very act of "writing with a pen," but now our technical world is gradually making us forget the actual essence of our day-to-day tasks. Lack of actual writing can eventually cause our bodies to develop paralysis towards using our hands in a proper way to write and can even sometimes lead to us forgetting the way of writing a particular language.

The process of digitalization, which is when digital technologies like AI and others are incorporated into numerous industries, has the power to completely alter how we live and work. The possible negative effects of such developments must be recognized and addressed in addition to the positive effects.

The prospect of job displacement is among the biggest worries associated with digitalization. There is a chance that some employees could become obsolete as

Data on employment shares and the proportion of jobs at high
risk of automation.

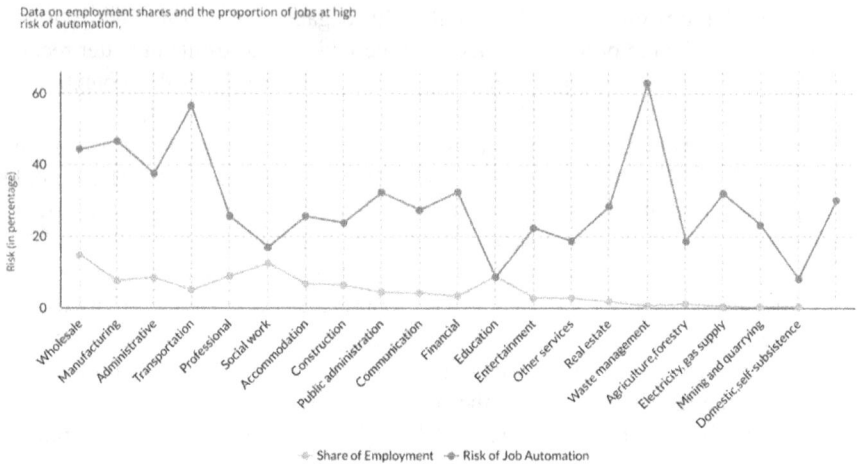

FIGURE 16.3   Data on employment shares and the proportion of jobs at high risk of automation.

more and more workers are mechanized using AI and other digital technologies. People who lack the expertise to work with cutting-edge technology may experience a major loss in employment chances as a result. Jobs that require repetitive work, like data input or customer service, for instance, are more likely to be automated. This might result in a sizable number of people losing their work, which might increase unemployment and poverty (Figure 16.3).

According to a new report, nearly half of all the work we do will be able to be automated by the year 2055. However, a variety of factors, including politics and public sentiment towards the technology, could push that back by as many as 20 years.

Here are the data representing the employment shares and job risk:

Another issue with digitalization is the potential for growing income disparity. The enhanced productivity and efficiency brought on by digitalization will be advantageous to those who have the abilities required to deal with cutting-edge technology. Without these abilities, people might have a hard time finding work and may fall behind in terms of income and style of living. This might exacerbate social and economic problems, resulting in a growing divide between the wealthy technocrats and the rest of society.

While digitalization has many advantages, it's vital to also take any potential disadvantages into account. To guarantee that people have the skills they need to adapt to the changing work market, proactive steps like education and training programs must be taken to address serious issues like job displacement and rising income disparity. To guarantee the appropriate and equitable use of digital technology, ethical and legal issues must also be addressed. As AI and digitalization continue to grow, these issues must be taken into consideration.

16.4   INTEGRATION OF DIGITALIZATION IN LEARNING

AI can help us in innumerable and unique ways, to solve daily life problems. What we can expect out of AI has humongous scope. Even today these technologies are being used, but the only thing is that it is not evenly distributed. Shortly, when the majority of the population will have access to these newer technologies, we will have to shift our focus from routine tasks to training our future generations to be much more efficient at decision making and creativity, because those will be the attributes for survival then.

Knowledge about these technologies and how to use them to promote creativity and enhance consumer experience is essential. For businesses to remain competitive in the market, it is also critical that they actively keep up with new technologies and look for methods to integrate them into their operations. Additionally, it is anticipated that shortly, technology will have a substantial impact on every industry and for this technology integration in the classroom is essential for students to be prepared for future, technology integration in the actual learning classroom is essential today. It is crucial to give today's learners, the knowledge and skills necessary to use technology, especially given the speed at which technology is developing and its growing influence across a wide range of industries. Beyond simply showing them how to use electronic gadgets, a technology-rich education system goes beyond that. It necessitates a more comprehensive curriculum that covers digital innovation, critical thinking, research, and problem solving. We have to use technology to engage students in the learning process. Incorporate technology into classroom activities and projects to help them visualize and explore content. The way pupils learn could be completely changed by the use of technology in the classroom and encouraging creativity and innovation. Inspire learners to think creatively and find innovative solutions to problems. Foster an environment where they can explore and experiment with technology. Offer them hands-on experience so as to make them ready to face the hi-tech world that would be double hi-tech by the moment they would step into their careers. Only computer usage won't suffice now; it cannot be the only aspect of teaching technology in schools now.

The curriculum should include a wide range of topics related to technology and how it affects their personal and professional destinies. For this, learners must possess the capacity to adjust to new technologies, keep up with technological developments, and continually develop their digital literacy. Only then learners can succeed in the constantly changing digital environment and bear the potential to contribute to its further development also.

16.5 CONCLUSION

The fields of AI and digitalization are both quickly developing and have the ability to fundamentally alter both corporate operations and societal advancement. But it's crucial to recognize and deal with any potential drawbacks of new technology, such as job loss, and wider wealth disparities. Investments in education and training programs are necessary to get people ready for future occupations and ensure that everyone in society may profit from these improvements. In addition, it's necessary to create laws and policies that encourage the retraining and upskilling of workers who may be replaced by automation. Additionally, wealth inequality needs to be addressed through social safety net programs and progressive taxation, as well as by redistributing income more fairly.

In addition, the ethical implications of this technology must be considered. These implications include worries about security, privacy, and possible biases in judgement. These dangers must be reduced, and the technology should be used sparingly, by making sure that accountability and openness are upheld during the design and deployment of AI systems. Additionally, it's also important to take into account how these technologies can influence underserved populations and to make sure that they don't suffer unduly by the negative effects of automation.

To guarantee that the technology is applied in a way that benefits society as a whole, the role of government in the regulation and oversight of AI development becomes far more significant. This may include rules governing the collection, usage, and sharing of data as well as standards for the creation and use of AI systems.

Overall, there is a lot of room for growth and advancement in the workplace, and society as a whole, as a result of the usage of AI and digitalization. We must adopt a proactive and all-encompassing approach to deal with any potential issues if we want to fully benefit from technology for everyone. In addition, to ensure that all stakeholders are included and have a voice in determining the direction and impact of new technologies, appropriate consideration should be given to the economic, ethical, and legal issues that may arise due to AI and digitalization.

16.6 FUTURE IMPLICATION OF DIGITALIZATION

Digitalization has a bright future ahead of it, one that has the power to change industries and enhance the quality of life for individuals all around the world. The speed of technological development, particularly in the areas of AI and ML, is enabling the automation of processes and the processing of massive volumes of data in ways that were previously unthinkable.

Digitalization holds great promise for the future and has the potential to significantly better people's lives. In a variety of industries, including healthcare, transportation, education, and banking, it is now possible to automate activities, process massive quantities of data, and make smarter judgements thanks to breakthroughs in AI and ML.

Interestingly enough, while winding up this research paper on digitalization we learned about a new bot in town ... Chat GPT. No doubt this paper will be incomplete if this essential and relatable invention is missing. An AI-based tool is threatening Google ... the father of all technical shifts. "Chat GPT is a big deal. The tool seems pretty knowledgeable in areas where there are good training data for it to learn from. It's not omniscient or smart enough to replace all humans yet, but it can be creative, and its answers can sound downright authoritative. A few days after its launch, more than a million people were trying out ChatGPT ... So ChatGPT, while imperfect, is doubtless showing the way toward our tech future," C Net.

Believe it or not, technology is in an exciting phase right now, and the opportunities are unlimited.

REFERENCE

Hembrooke, H., & Gay, G. (2003). The laptop and the lecture: The effects of multitasking in learning environments. *Journal of Computing in Higher Education, 15,* 46–64.

Index

Note: **Bold** page numbers refer to tables and italic page numbers refer to figures.

For Product Safety Concerns and Information please contact our EU
representative GPSR@taylorandfrancis.com
Taylor & Francis Verlag GmbH, Kaufingerstraße 24, 80331 München, Germany

www.ingramcontent.com/pod-product-compliance
Lightning Source LLC
Chambersburg PA
CBHW060553220326
41598CB00024B/3095